Smart Agriculture

Volume 9

The book series Smart Agriculture presents progress of smart agricultural technologies, which includes, but not limited to, specialty crop harvest robotics, UAV technologies for row crops, innovative IoT applications in plant factories, and big data for optimizing production process. It includes both theoretical study and practical applications, with emphasis on systematic studies. AI technologies in agricultural productions will be emphasized, consisting of innovative algorithms and new application domains. Additionally, new crops are emerging, such as hemp in U.S., and covered as well. This book series would cover regions worldwide, such as U.S., Canada, China, Japan, Korea, and Brazil.

The book series Smart Agriculture aims to provide an academic platform for interdisciplinary researchers to provide their state-of-the-art technologies related to smart agriculture. Researchers of different academic backgrounds are encouraged to contribute to the book, such as agriculture engineers, breeders, horticulturist, agronomist, and plant pathologists. The series would target a very broad audience – all having a professional related to agriculture production. It also could be used as textbooks for graduate students.

Yuliang Yun · Wenyi Sheng · Zhao Zhang
Editors

Advanced Sensing and Robotics Technologies in Smart Agriculture

Springer

Editors
Yuliang Yun
College of Mechanical and Electrical Engineering
Qingdao Agricultural University
Qingdao, Shandong, China

Wenyi Sheng
Research Center for Smart Agriculture
China Agricultural University
Beijing, China

Zhao Zhang
Key Laboratory of Smart Agriculture System Integration Research, Ministry of Education
China Agricultural University
Beijing, China

ISSN 2731-3476 ISSN 2731-3484 (electronic)
Smart Agriculture
ISBN 978-981-97-6440-2 ISBN 978-981-97-6441-9 (eBook)
https://doi.org/10.1007/978-981-97-6441-9

This Springer imprint is published by the registered company Springer Nature Singapore Pte Ltd.
The registered company address is: 152 Beach Road, #21-01/04 Gateway East, Singapore 189721, Singapore

Preface

Conventional agricultural approach faces challenges to meet the requirements of continuously growing population, and simultaneously, it is not in a sustainable manner. In addition, the increasing labor cost and shrinking labor pool threaten the agriculture development. Smart agriculture adopts advanced technologies and data-driven production operations to improve the agricultural production. Though the smart agricultural technology concept has been proposed decades ago, it does not come to practical applications until very recently, because of recent progress in advancing sensing, artificial intelligence, internet of things, and automation. With rapid technology progress in smart agricultural technology, it is required to propose a book to summarize the progress comprehensively and systematically.

The current book aims to present the recent technology progress in smart agriculture area, with former five chapters focusing on reviewing innovative technologies and the latter two chapters on honeysuckle detection and wheat lodging monitor. Seed quality plays an important role in guaranteeing good yield, and Chap. 1 works on detecting seed quality using appearance characteristics. After the seed germination, it is important to sensing the growing status of the crops, which generates the development of Chap. 2. Chapter 3 then specifically pays attention to greenhouse crop phenotyping, followed by Chap. 4 on tomato and apple on-site intelligent grading systems. The last review chapter works on reviewing harvest robotics. Following Chap. 5 review chapters, Chap. 6 improves YOLOv5s for infield honeysuckle detection under natural lighting, and Chap. 7 monitors wheat lodging conditions using improved MobileNetV3. To sum up, the seven chapters provide a general idea on the smart agricultural technology progress in recent years.

Drs. Yuliang Yun, Wenyi Sheng, and Zhao Zhang, the editors of this book titled *Advanced Sensing and Robotics Technologies Applied in Smart Agriculture—State-of-the-Art Technologies* have been working in the smart agriculture area for decades, and their expertise, knowledge, research experience and practical applications guarantee the high-quality content of this book. This book provides a comprehensive and timely information source for readers who are interested in learning about the important subject of smart agriculture.

Dr. Yuliang Yun
Associate Professor of College of Mechanical and Electrical Engineering
Qingdao Agricultural University
Qingdao, China

Dr. Wenyi Sheng
Associate Professor of College of Information and Electrical Engineering
China Agricultural University
Beijing, China

Dr. Zhao Zhang
Professor of College of Information and Electrical Engineering
China Agricultural University
Beijing, China

Contents

Chapter 1
Research Progress on Seed Appearance Recognition for Major Crops

Yuliang Yun, Dehao Li, Xueke An, and Zhiyong Ma

Abstract The appearance characteristics of crop seeds can provide information about the quality, size, and variety of seeds. Through continuous research on seed appearance recognition technology, automated grading and screening of large-scale seed samples can be realized to ensure the purity and consistency of seeds and improve the quality and efficiency of crop production and breeding. Therefore, strict quality testing and identification of crop seeds are key factors in ensuring high yield, high quality, and crop safety. Currently, the main categories of appearance recognition technology include manual recognition, machine vision technology, and spectral imaging technology. Machine vision technology simulates the visual function of the human eye and combines image processing and analysis algorithms to achieve rapid and accurate identification of seed appearance characteristics. Spectral imaging technology, specifically near-infrared spectroscopy (NIR), analyzes the spectral information on the surface of seeds to reveal their internal chemical composition and structural characteristics, enabling non-destructive detection of seed varieties and quality. The application of machine vision recognition technology and spectral imaging technology provides a new solution for crop seed appearance recognition, improving the accuracy and efficiency of seed recognition and promoting the optimization of the agricultural production chain and the competitiveness of agricultural products in the market.

Keywords Crop seed · Appearance recognition · Machine vision · Spectral imaging

Y. Yun (✉) · D. Li · X. An · Z. Ma
College of Mechanical and Electrical Engineering, Qingdao Agricultural University, Qingdao 266109, China
e-mail: ylyun1981@163.com

Y. Yun et al. (eds.), *Advanced Sensing and Robotics Technologies in Smart Agriculture*, Smart Agriculture 9, https://doi.org/10.1007/978-981-97-6441-9_1

1.1 Introduction

As the global population continues to grow and urbanization accelerates, food security has become one of the major challenges facing the world. As the core of agricultural production, the quality of seeds is directly related to the growth and development, yield, and quality of crops, which in turn affects the stability and security of the entire food supply chain. Therefore, improving the quality of seeds and ensuring their health and genetic superiority plays a crucial role in enhancing the efficiency of agricultural production, guaranteeing food security, and realizing sustainable development.

High-quality seeds are a good starting point for crop growth. In agricultural production, the main crop seeds include rice, wheat, peanuts, corn, cotton, soybeans, and other seeds, as shown in Fig. 1.1. In this paper, we will also delve into the appearance recognition techniques applied to these six types of crop seeds. Crop seeds are not only an important part of global food production but also play a central role in guaranteeing global food security and promoting the process of agricultural modernization. By exploring the current development and future trends of crop seed appearance recognition technology, we will gain a deeper understanding of the needs for seed quality assessment and provide strong support for the sustainable development of global agricultural production.

As the foundation of crop production, seed quality testing and accurate variety identification are of great significance in achieving high-yield, high-quality, and safe crop production. The quality of seeds is directly linked to crop growth and harvest yield. High-quality seeds ensure optimal germination and growth potential, enhance crop resistance to pests and diseases, and reduce the need for pesticides

Fig. 1.1 Seed of major crops: (**a**) Rice seed; (**b**) Wheat seed; (**c**) Peanut seed; (**d**) Corn seed; (**e**) Cotton seed; (**f**) Soybean seed

and other resources that are harmful to crops. This effectively improves crop yield and quality. Thus, quality control starting from seeds is crucial for optimizing the entire agricultural production chain and enhancing the market competitiveness of agricultural products.

Accurate identification and quality screening of seeds form the basis for selecting superior seeds for breeding and management in agricultural operations. With the rapid development of information technology and biotechnology, seed appearance recognition technologies have become a vital research area in modern agriculture. These technologies not only enhance the accuracy of seed screening and classification but also significantly increase processing speed and reduce labor costs. Variety identification of seeds also provides essential data and samples for agricultural scientific research, supporting the conservation and utilization of crop genetic breeding and genetic resources. Furthermore, through precise seed appearance recognition, early detection of seed diseases and defects is possible, enabling effective quality control and risk management prior to planting, thereby ensuring overall crop quality and yield.

The development and application of seed appearance recognition technology brings great potential and opportunities for agricultural production. By using machine vision and image processing technology, features such as shape, texture and color of seeds can be quickly and accurately extracted and analyzed to achieve automated identification and classification of seeds; in addition, spectral imaging technology is also a key technology for seed appearance recognition, which achieves high-precision identification and classification of seed appearance by measuring and analyzing the spectral features of seeds at different wavelengths. Different seeds present different features in absorption, reflection and transmission spectra, including color, texture, chemical composition, etc. Through spectral imaging technology, a large amount of spectral data can be acquired, thus obtaining more comprehensive and accurate information on seed appearance.

Traditional seed identification methods rely heavily on human expertise and visual inspection by relying primarily on manual visual inspection and manual categorization [1]. This approach suffers from the disadvantages of inefficient identification, subjectivity, and high labor costs. The need to improve the accuracy and efficiency of seed quality assessment has driven the development of various technological solutions aimed at automating these processes. The comparison between appearance recognition technology and traditional manual recognition under different features is shown in Table 1.1.

The introduction of machine vision systems in the early 2000s marked a major shift in seed technology. As detailed by Patel and Kumar [2], machine vision employs cameras and image processing algorithms to analyze physical characteristics of seeds, such as shape, size, and color. These systems provide a faster and more consistent alternative to manual inspection. For example, Martin and Davis [3] emphasized the capability of advanced machine vision systems to conduct high-throughput and high-precision seed sorting, greatly enhancing the productivity of seed processing operations.

Table 1.1 Comparison of seed appearance inspection techniques

Features	Seed Appearance Recognition Technology	Traditional manual identification of seed appearance
Automatization	Automatic extraction and recognition of seed features using machine vision or hyperspectral techniques	Reliance on manual visual inspection and judgment
Accuracy	High accuracy with consistency and reliability	Subjective factors interfering, large errors
Speed	Fast processing of large numbers of seed samples, real-time or high-throughput assays	Relatively slow, requiring more time and human resources
Data analysis and recording	Quantitative analysis, digital records and statistics	Subjective description, difficult to analyze quantitatively
Scaleability	Scalable to handle large-scale seed samples	Difficult to scale up to large-scale seed sample testing needs
Automatic defect detection	Can automatically detect and recognize seed defects, diseases, and mutations	Lack of automated defect detection capabilities and reliance on specialized knowledge and experience

Parallel to the development of machine vision, neural networks and deep learning algorithms are gaining increasing prominence in the field of seed technology. Lee and Park reviewed deep learning models developed for seed quality assessment [4], emphasizing their capacity to learn complex patterns in data, which may be overlooked by traditional machine learning methods. Studies conducted by Gomez and Lee demonstrated the efficiency of convolutional neural networks (CNNs) for seed recognition, showcasing their superior accuracy in distinguishing highly similar seed varieties based on subtle visual cues [5].

Another significant technological advancement is the utilization of spectral imaging, which offers a non-destructive approach to assess seed quality. Spectral imaging captures data at multiple wavelengths, enabling a comprehensive understanding of the chemical and physical properties of seeds [2]. This method proves particularly valuable in detecting internal defects or contaminants that are not visible to the naked eye. Nelson conducted research on the application of electromagnetic and optical spectroscopy in quality assessment, demonstrating how these techniques can effectively enhance the seed sorting process by providing detailed insights that surpass visual analysis [6].

The integration of these advanced technologies has not only improved the accuracy and efficiency of seed quality assessment, but has also facilitated the development of high-throughput systems capable of rapidly processing large quantities of seeds. The systems discussed by Larson and White and Wu and Sun have now become integral parts of the seed industry, accelerating decision-making processes and significantly reducing the amount of labor required for seed sorting and classification [7, 8].

In addition, the development of seed technology has been influenced by the increasing emphasis on sustainable agricultural practices. Advanced seed sorting technologies help reduce waste by accurately identifying viable and non-viable seeds, ensuring that resources are not wasted on seeds with poor germination prospects. This is critical in an era where the environmental impact of agricultural practices is under scrutiny.

In summary, the field of seed technology has made significant advances over the past two decades, driven by the integration of machine vision, deep learning, and spectral imaging. These technologies have transformed seed quality assessment from a labor-intensive, error-prone process to a more reliable, efficient, and automated operation. As these technologies continue to evolve, they are expected to further improve the accuracy and efficiency of seed sorting and classification, supporting sustainable agricultural practices and meeting the growing global demand for food.

1.2 Basic Features of Crop Seed Appearance

Seed quality is critical to successful agriculture because it directly affects germination rates, plant health, and crop yields. Key attributes of seed appearance include size, shape, color, and texture. Each of these characteristics can provide valuable insights into seed health and the potential for successful cultivation. In this context, understanding the relationship between seed appearance and quality is crucial for optimizing agricultural outcomes.

Seed morphology description:

SIZE: Seed size is a fundamental characteristic that is often associated with seed vigor and energy reserves. Larger seeds typically contain more resources to support early growth stages, resulting in more robust seedlings. Technologies such as machine vision systems have been used to accurately and consistently measure seed size, which can aid in the sorting process for classifying seeds based on their potential vigor [8].

Shape: Seed shape can vary greatly between species and is often a key determinant of species identification. Shape also affects the mechanical seeding ability of seeds and thus their suitability for modern agricultural practices. Figure 1.2 shows several types of peanut seeds with different shapes. Automated sorting systems using image analysis are capable of differentiating seeds based on geometric parameters, ensuring that irregularly shaped seeds that do not meet the requirements of a particular planting device are screened out [9]. Dana et al. investigated an image description method for the shape and color of flax seeds, extracting four shape features and three color features for seed classification. Principal component analysis and the unweighted

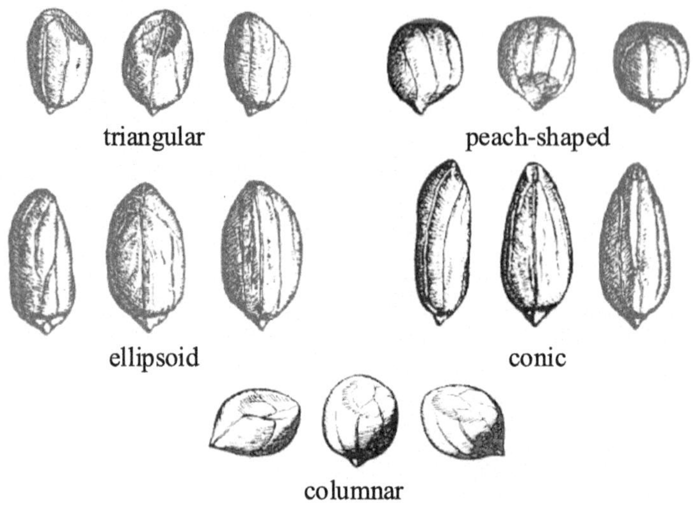

Fig. 1.2 Peanut seed shapes

group averaging method (pair group method with arithmetic mean (UPGMA)) were used for qualitative cluster analysis [10].

Color: The color of seeds can indicate their maturity, type, and sometimes even exposure to diseases or pests. For instance, discoloration may suggest a fungal infection or improper storage conditions, which can compromise seed quality. Advanced imaging techniques, such as hyperspectral imaging, go beyond visible color analysis by capturing data at multiple wavelengths to detect subtle color variations that indicate seed health [6]. Yan et al. extracted the color information of crown and side images of maize seeds under RGB and HSV color models [11], and discriminated them using the Fisher discrimination method, with correct recognition rates of different varieties of maize seeds of up to 1.5 times higher than those of other varieties. The correct recognition rate for different varieties of corn seeds was 93.75%. At present, the mainstream peanut seed screening equipment is mainly the traditional mechanical screening machine and the color sorter according to the appearance of the crop seed color photoelectric color selection. Autoline is a company specializing in the production of optical classifiers, its optical classifiers can be agricultural products, food, pharmaceuticals, etc. for accurate grading, screening and classification. The United States of America's photoelectric color sorter are used in Japan Satake Co., Ltd. production of equipment, in addition to Satake and Switzerland's Buhler, South Korea's Daehara, Norway's TOMRA and other companies are engaged in agricultural products, color sorter of the relevant research and development. The device diagrams are shown in Fig. 1.3.

Texture: The surface texture of seeds can exhibit characteristics such as smoothness, roughness, or the presence of hairs or other structures. These texture features can impact seed-soil interactions, moisture absorption, and even susceptibility to

Autoline Optical Classifier, USA

Japan Satake Machinery EZS Series Sorting Machine

Buhler, Switzerland SORTEX F BioVion color sorter

Fig. 1.3 Related color sorting machines

diseases. Machine vision and deep learning algorithms are particularly effective in analyzing these complex patterns on seed surfaces, providing detailed assessments that can predict quality based on texture features [5]. In 2008, Manickavasagan et al. utilized a monochromatic image system to capture grayscale images of eight different species of wheat. They extracted 32 texture features and achieved correct recognition using linear discriminant and quadratic discriminant analysis, with recognition rates ranging from 73 to 100% [12].

Seed appearance not only aids in seed identification and classification but also serves as a visual indicator of seed health and quality. High-quality seeds typically demonstrate uniformity in size, shape, and color, and are devoid of physical defects and disease symptoms.

Health assessment: Visual inspection of seeds can unveil signs of diseases, insect infestation, or physical damage. For instance, seeds exhibiting unusual spots or color changes may be infected or may have been stored under deteriorating conditions. Advanced imaging techniques, such as spectral imaging, enable more comprehensive inspection by detecting chemical changes in seeds that are not visible to the naked eye [2].

Quality prediction: The overall appearance of a seed can serve as a reliable predictor of its germination ability and potential yield. Automated systems equipped with artificial intelligence can swiftly analyze large quantities of seeds, evaluate their appearance based on predetermined quality parameters, and effectively predict their vigor [4].

Sorting and categorization: Automated sorting systems utilize attributes of seed appearance to differentiate high-quality seeds from low-quality ones. This sorting process ensures that only the best seeds are planted, which is crucial for maximizing crop yield and quality. Such systems have become integral to modern agriculture, where accuracy and efficiency play vital roles in productivity [3].

The appearance characteristics of crop seeds are not only influenced by genetic and environmental factors but also serve as a crucial basis for assessing seed quality in agricultural production. By accurately analyzing these characteristics, the efficiency of seed selection and classification can be significantly improved, ensuring that healthy and high-growth-potential seeds are chosen for planting. Moreover, these appearance indicators are also linked to food safety and the ultimate quality of agricultural products. High-quality seeds are more likely to yield high-quality crops, thereby enhancing agricultural output and market competitiveness. With technological advancements, particularly in the fields of image processing and data analysis, a more precise assessment and application of seed appearance characteristics can be achieved, leading to efficient and sustainable agricultural production. This optimization of resource utilization and reduction in agricultural production costs will also contribute to environmental protection and promote the development of agriculture in a more modern and automated direction.

1.3 Methods of Recognizing the Appearance of Crop Seeds

At present, the recognition methods for crop seed appearance can be categorized into traditional manual recognition, machine vision analysis technology, and spectral imaging analysis technology. Manual recognition is a commonly employed method for seed appearance assessment. It relies on human experience and expertise to classify and evaluate seeds by directly observing and recognizing their appearance characteristics. Machine vision utilizes computer vision algorithms and image processing techniques for automated identification. It enables computer systems to understand and analyze image or video data by capturing seed images, extracting image features such as color, shape, and texture, and employing machine learning algorithms for classification and identification. Deep learning and neural networks are integral components of machine vision. Spectral imaging analysis techniques, on the other hand, are non-destructive and rapid. Hyperspectral techniques, in particular, enable the acquisition of a large amount of spectral information, allowing for more accurate capture and analysis of seed appearance features.

1.3.1 Manual Identification Methods

In crop seed appearance identification, manual identification relies mainly on the experience and intuition of the observer to determine the quality and type of seed. However, manual identification methods have shown significant limitations in the context of modern agricultural technology, particularly in terms of processing efficiency and accuracy. This method heavily relies on the subjective judgment and experience of the operator, resulting in inefficiencies when processing seeds on a large scale, as well as poor consistency and reproducibility of results. Moreover, manual identification is susceptible to individual fatigue and distraction, which can lead to increased misjudgments, ultimately affecting the quality of the final selection, as shown in Fig. 1.4. These factors render manual recognition methods unsuitable for modern seed processing requirements, necessitating the urgent need to enhance automation, accuracy, and processing speed through modern techniques such as machine vision recognition technology and spectral imaging technology.

Fig. 1.4 Manual seed sorting

1.3.2 Machine Vision Recognition Technology

Machine vision is a field of artificial intelligence that involves enabling computers and systems to extract valuable information from images, videos, and other visual inputs, and to take actions or provide recommendations based on that information. Common tasks in machine vision include image classification, object detection, object tracking, and content-based image retrieval. To accomplish these tasks, machine vision relies on techniques such as image processing and pattern recognition, including image filtering, edge detection, feature extraction, target detection, image classification, and deep learning.

1.3.2.1 Principle of Machine Vision Technology

Machine vision technology is a technique for recognizing, classifying, and locating target objects by using cameras, image sensors, and digital processing systems to simulate the visual perception of the human eye. The development of this technology stems from the need for automation and precise control, especially in a wide range of applications such as industrial manufacturing, agriculture, medical diagnostics, and many other fields.

The basic workflow of a machine vision system includes steps such as image acquisition, image preprocessing, feature extraction, recognition and decision making. The machine vision inspection process is shown in Fig. 1.5.

Image acquisition: First, images of the target object are captured by means of cameras or other image-capturing devices. These devices must be able to ensure image quality under various environmental conditions for subsequent processing.

Image pre-processing: The purpose of image pre-processing is to improve image quality for subsequent processing. This includes noise removal, contrast adjustment, brightness adjustment, etc. Preprocessing may also include image normalization, such as normalizing image size and color depth.

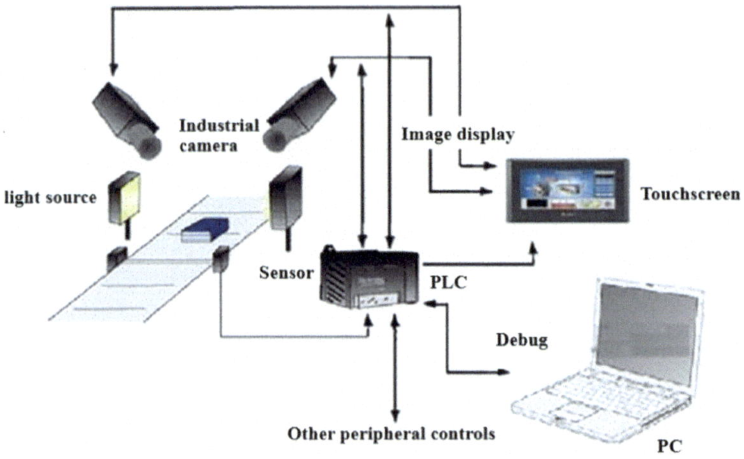

Fig. 1.5 Machine vision inspection process diagram

Feature extraction: extracting useful information from the preprocessed image, such as edges, corner points, texture, etc. These features should be representative and distinguishable to effectively support subsequent image analysis.

Recognition and Decision Making: Finally, based on the extracted features, classification algorithms (e.g., Support Vector Machines, Decision Trees, Neural Networks, etc.) are used to recognize targets in the image. This step usually involves complex algorithms and a large number of computations, and is the most critical part of the whole machine vision system. This paper studies the appearance recognition technology of major crop seeds. The process of crop seed appearance feature extraction is shown in Fig. 1.6.

As technology evolves further, machine vision systems are beginning to merge with other technologies such as robotics, the Internet of Things, and edge computing to form smarter, more automated solutions. For example, the combination of machine vision and robotics allows robots to perform more precise operations in complex environments, such as automated picking and assembly. In addition, with the popularity of IoT devices, machine vision systems are being integrated into security monitoring in smart homes and smart cities to provide real-time image analysis.

1.3.2.2 Machine Vision Recognition Technology Development and Application

The development of machine vision commenced in the 1950s, coinciding with the emergence and progress of computer science. Initially, these techniques found applications primarily in the medical and military domains, as the imaging needs of these fields drove the advancement of related technologies. From the 1970s to the 1990s, with the improvement in computer processing power and the advancements in digital

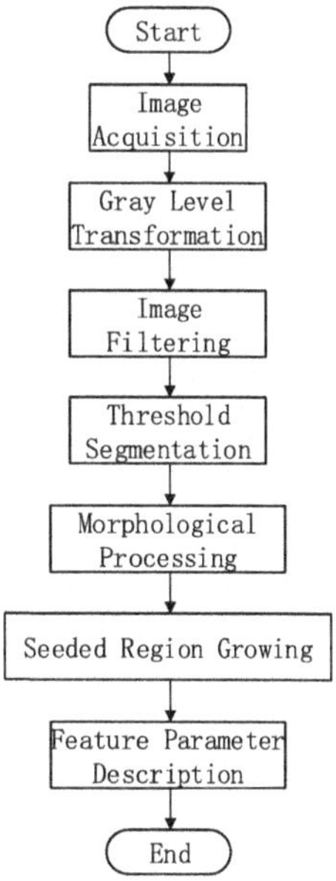

Fig. 1.6 Process of crop seed appearance feature extraction

image processing technology, machine vision gradually entered industrial applications. It found particular use in automated production lines for tasks such as quality inspection, object recognition, and assembly control.

In the early twenty-first century, the rise of artificial intelligence and machine learning, particularly the development of deep learning, brought about a significant breakthrough in machine vision technology. Deep learning models such as convolutional neural networks (CNN) demonstrated remarkable performance in tasks like image classification, target detection, and image recognition. This advancement greatly expanded the application areas of machine vision technology. During this period, machine vision systems were not limited to traditional industrial manufacturing fields but also found widespread use in emerging areas such as autonomous driving, intelligent surveillance, agricultural automation, and medical image analysis.

As machine vision continues to evolve, deep learning and neural network technologies are continuously updated and iterated upon. They have been successfully applied in agriculture, offering substantial advantages over traditional machine learning methods for crop seed appearance recognition and quality detection. Many researchers in the field of agriculture have utilized machine vision technology to address production challenges, particularly in major crops like peanuts, corn, soybeans, wheat, rice, and more, leading to significant advancements.

For example, as early as 1986, Gunasekaran et al. proposed a crack defect detection method to identify corn kernels [13], showing that corn kernel cracks and other parts of the high-speed filtering method can be used to identify them, the detection accuracy of up to 90%.Neuman et al. in 1989, using the R, G, B value of the wheat kernel is divided into six grades of categories [14]. A tube color camera was used for image acquisition, and the average R, G, B pixel reflectance values of each wheat kernel were used as color attributes to distinguish feature differences between different wheats. Liao. K et al. in 1992 classified maize kernel images into five classes of kernel morphology by using 34 feature parameters as the input variables of a neural network. The learned neural network classified the kernels with an accuracy of 93 and 91% accuracy for classification of broken kernels [15]. Chtioui et al. in 1999 proposed the use of Rough sets theory as a pattern classification tool for evaluating the characteristics of broad beans by computer vision techniques. The theory proposes to use two different discrete methods to differentiate between broad beans with different characteristics and impurities, and use 35 feature parameters extracted from color images to classify them, and the classification results have a good degree of consistency compared with the statistical classification results of discriminant analysis [16]. Hecang et al. set up a peanut high-throughput automatic detection system, which can measure the parameters of peanut pods and nuts in real time, and the length, width, diameter, and number of peanut pods and seeds The average measurement accuracy is more than 98% [17]. Balaji et al. proposed DeepSort deep convolutional network for sorting haploid maize seeds, and experimentally demonstrated the possibility of DeepSort to distinguish haploid and diploid in the process of double haploid induction in maize seeds [18]. Kurtulmuş investigated the three AlexNet, GoogleNet and ResNet deep learning architectures to recognize sunflower seeds, and the experimental results showed that the GoogleNet algorithm had up to 95% accuracy [19]. Shima et al. proposed a new method using deep convolutional neural network as a generalized extractor, and compared with multiple classifiers, the CNN-ANN classifier performed best in the corn seed variety classification test with 98.1% accuracy [20]. Nandin Sidnal designed a classification system using image processing and PNN to differentiate grains with 80–90% accuracy [21]. Guoyang Zhao et al. designed and developed a soybean seed sorting system based on the improved MobileNet V2 model, which uses an alternating rotation mechanism to display the complete surface feature information of the seeds, and achieves real-time identification of the whole surface of soybeans with an inference speed of 35 fps on NVIDIA's Jetson Nano board [22]. Chang Song et al. proposed a seed quality assessment method using an improved Inception-ResNet network to identify corn seeds of different qualities. The average accuracy was 96.27%, the average recall was 96.03%, the F1 value was 96.15%, and the detection speed of a single corn seed image was about 2.44s [23]. Ouyang et al. proposed a method for automatic identification of rice varieties by machine vision technology, and developed an inspection system consisting of an automatic inspection machine and an image processing unit. A dedicated rice variety detection software was developed [24]. C Bi et al. proposed a neural network corn seed image classification model MFSwin transformer based on feature attention and multi-scale feature fusion. The average recognition accuracy is 96%, and the number of parameters is only half of the original model [25]. Liu Zhicheng et al. proposed an algorithm based on improved YOLOv5 target detection of cottonseed, by adding lightweight upsampling operator CARAFE, CIoU is replaced by EIoU, and the test results show that the improved algorithm achieves a mAP_0.5 value of 99.2%, and the recall rate reaches 98.9% [26]. Xuexing et al.

proposed a convolutional neural network-based image recognition method for wheat seeds, GC_DRNet, by changing the residual module to a dense residual module and introducing a global context module, the recognition accuracy of the model was improved by 2.34% and the number of parameters was reduced by 73.87% compared with the original ResNet18 network model [27].

1.3.3 *Spectral Imaging Technology*

Spectral imaging technology, also known as hyperspectral imaging or spectral image technology, is an advanced imaging technique that combines the benefits of traditional imaging and spectral analysis to capture both spatial and spectral information of target objects. By analyzing the spectral data of each pixel point, this technology has found widespread application in various fields, including agriculture, environmental monitoring, and medical diagnosis.

1.3.3.1 Principles and Equipment for Spectral Imaging

Spectral imaging techniques are based on the spectral characteristics of light reflected or transmitted from an object. Each substance absorbs light at specific wavelengths, and these absorption features are characterized by specific "fingerprints" on the spectrum. By analyzing these spectral fingerprints, substances can be identified and classified.

Spectral imaging can be divided into four steps: optical acquisition, spectral separation, spectral data recording, image reconstruction and analysis.

The spectral imaging system consists of three main components:

*Optical system: including various lenses and optical filters for focusing and directing light.

*Spectroscopic elements: such as prisms or diffraction gratings, which are used to decompose light into its constituent spectra.

*Detector: usually a photodetector, such as a CCD or CMOS sensor, which is used to record the intensity of light at different wavelengths to produce an image.

Figure 1.7 shows two common types of hyperspectral cameras and spectrometers. Figure 1.8 shows the schematic structure of a typical hyperspectral imaging system.

The principle of spectral imaging technology is shown in Fig. 1.9. Spectral imaging systems begin by capturing an image of a target object through an optical device such as a camera. In this process, light of different wavelengths strikes the object, which absorbs and reflects light of different wavelengths according to its chemical and physical properties. The captured light is separated by spectroscopic elements such as prisms or diffraction gratings. These devices are capable of breaking down the light into the different wavelengths of which it is composed. In this step, the raw optical signal is converted into spectral data at a series of wavelengths, each

Fig. 1.7 Specim FX10 line-scan hyperspectral camera with NIR25S near-infrared spectrometer

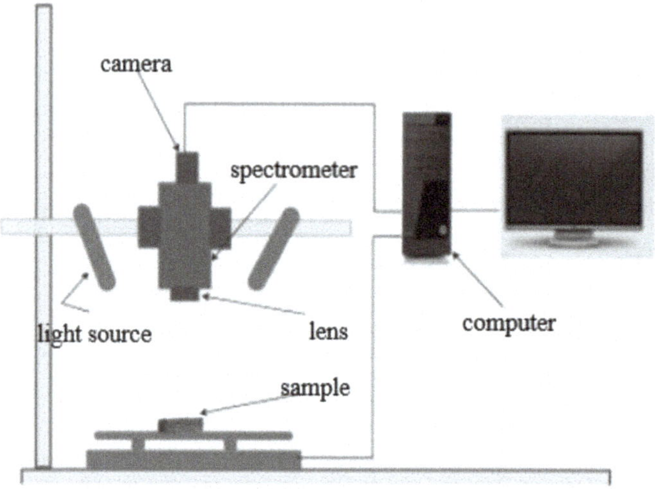

Fig. 1.8 Spectral imaging detection process diagram

corresponding to a different spectral intensity. The separated spectrum is sent to a detector (usually a CCD or CMOS sensor) where it is converted into an electrical signal. The detector records the intensity of the spectrum at each particular wavelength, generating digital data representing the spectral intensity. This data can be displayed as a spectrogram, with each pixel containing a complete spectral profile. Finally, processing software combines and reconstructs this spectral data into an image where each pixel point contains not only color information but also detailed spectral information for that point. By analyzing these images, different substances in the image can be identified and classified.

1.3.3.2 The Application of Near-Infrared (NIR) Spectroscopy in Seed Appearance Identification

Hyperspectral technology is rapidly advancing and has been extensively studied in the military, geological, and agricultural fields. In the realm of crop seeds, researchers have utilized hyperspectral technology to investigate various qualities ranging from surface characteristics to internal properties of seeds. Hyperspectral images can

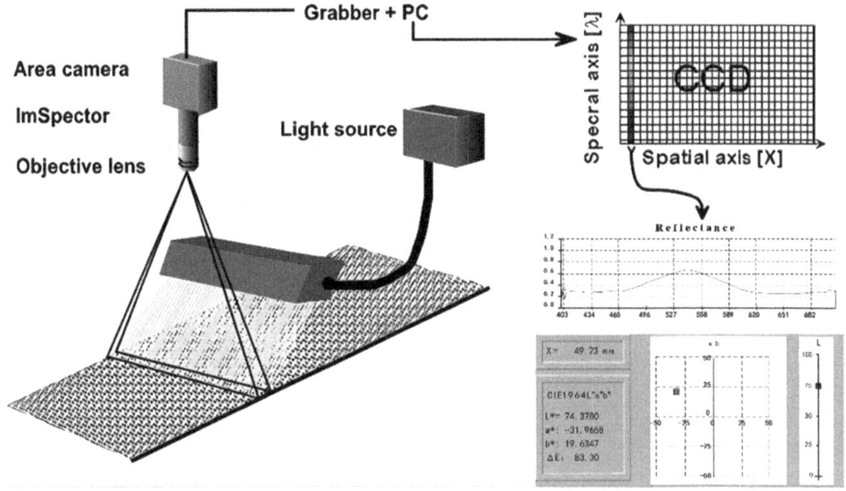

Fig. 1.9 Spectral imaging technology schematic diagram

capture the spectral information of each part of the seed, enabling further analysis. The extraction and classification process of hyperspectral image data is shown in Fig. 1.10. This addresses the issue encountered in analyzing a single seed with near-infrared spectroscopy (NIRS), where the collected spectra may not reflect the specific characteristics of different parts of the seed due to the heterogeneous distribution of seed components.

Additionally, the morphological and textural features of seeds can be extracted by utilizing the image information from each band in the hyperspectral image. Therefore, the use of hyperspectral technology for assessing crop seed quality combines the advantages of image detection and spectral detection, allowing for separate analysis of different seed parts.

NIRS, with a wavelength range of 750–2500 nm, has been widely employed as a fast, nondestructive, and efficient detection method in various fields such as petroleum, chemical industry, agriculture, and food. Since Norris et al. first used NIRS to analyze water content in agricultural products, researchers have conducted extensive studies on its application in agricultural and food quality testing [28].

In crop production, many diseases can be transmitted through seeds, making rapid detection of disease infection in seed appearance crucial for agricultural production. Dewiche et al. used near-infrared reflectance spectroscopy in the 940–1700 nm range to detect mildew and downy mildew in wheat. They explored the feasibility of using a small number of wavelengths or dual wavelengths to classify wheat, employing an exhaustive search method to find the optimal combination of single-wavelength algorithms. The study identified 1182 nm and 1242 nm as the best bands for detecting wheat mold and mildew, providing a theoretical basis for the technique's application in wheat classification and commercial sorting [29].

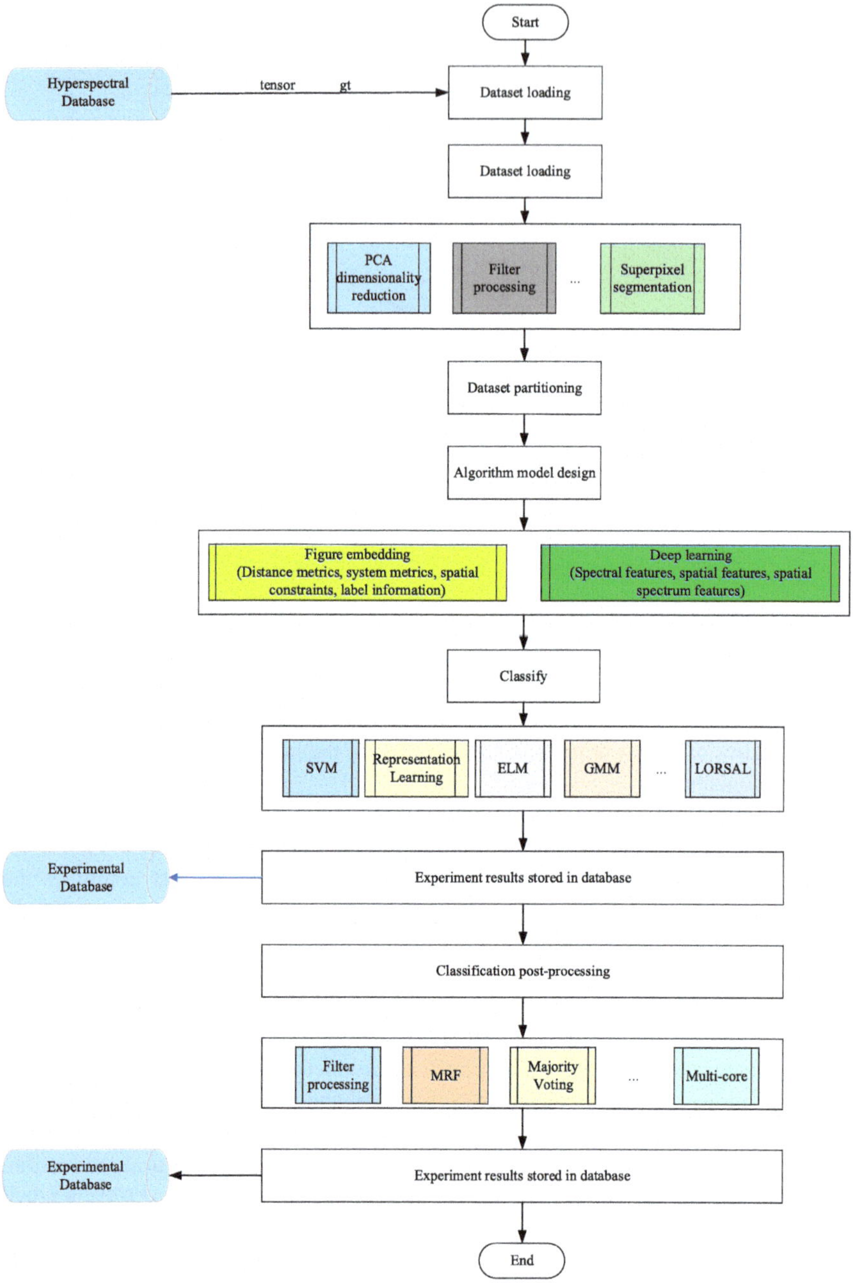

Fig. 1.10 Process of hyperspectral image data extraction and classification

As early as 2007, Ph. Vermeulen et al. proposed a fast and reliable method to identify barley varieties by near-infrared hyperspectral imaging [30]. Yanghuan et al. proposed a method for maize seed variety identification using stochastic subspace integrated learning and hyperspectral imaging. In the wavelength range of 400–1000 nm, in order to improve the identification accuracy of maize seed varieties, a stochastic subspace integrated learning (RSEL) model was established using LDA as the base classifier. Using MSC-IRIV-RSEL (Multiplicative Scattering Correction-Iterative Retention of Information Variables-Random Subspace Integrated Learning), the accuracy was improved from 0.9333 to 0.9556 and the Kappa coefficient was improved from 0.9174 to 0.9457 [31]. Zhu Shao-Long et al. collected hyperspectral images of 10 soybean seed varieties and obtained reflectance. The results showed that MSC-CARS-EL (Multiplicative Scattering Correction-Competitive Adaptive Re-weighted Sampling-Integrated Learning) had the highest accuracy, with 100%, 100%, and 99.8% accuracies for the training set, the test set, and the fivefold cross-validation, respectively [32]. Baichuan Jin et al. used near-infrared (NIR) hyperspectral technology combined with conventional machine learning methods (support vector machine (SVM), logistic regression (LR), and random forest (RF)) and deep learning methods (LeNet, GoogLeNet, and residual network (ResNet)) to build a variety identification model for five common rice seeds [33]. Zhang Han et al. used hyperspectral imaging to collect the transmittance and reflectance spectra of wheat seeds and used partial least squares discriminant analysis (PLS-DA) combined with various preprocessing methods to establish a classification model. The Detrend-CARS-PLS-DA model was established using transmittance spectra combined with a characteristic wavelength screening algorithm, with accuracies of 95.69, 98.25, and 97.25% in the test sets of King Pulse 9, King Pulse 11, and King Pulse 183, respectively [34].

Ashabahebwa Ambrose et al. investigated the possibility of using hyperspectral imaging (HSI) techniques to differentiate between viable and nonviable maize seeds. Partial Least Squares Discriminant Analysis (PLS-DA) was developed to classify aged (heat treated) and normal (untreated) maize seeds. The model showed the highest classification accuracies of 97.6% (calibration) and 95.6% (prediction) in the SWIR region of HSI [35]. A method based on information fusion and Artificial Fish Swarm Algorithm (AFSA) combined with Hyperspectral Imaging (HSI) of five rice seeds was proposed by Jun Sun et al. [36]. The study showed that the performance of the SVM model based on fused features was significantly improved, with a model accuracy of 97.22%. After AFSA optimization, the model accuracy finally reached 99.44%. Zhu Qibing et al. extracted the image entropy of hyperspectral images as the classification features of 17 maize varieties, including 1632 samples, with the spectral region covering 400–1000 nm and containing 233 wavelengths. The 65 best wavelengths were selected using the partial least squares (PLS) projection algorithm. Partial Least Squares Discriminant Analysis (PLSDA) was used to build a classification model for corn seed purity. The results show that the training accuracy of the model at the optimal wavelength (only 27.90% of the full wavelength) is 99.19%, and the testing accuracy is 98.90% [37]. Qiu Zhengjun et al. investigated the feasibility of utilizing convolutional neural network (CNN) hyperspectral imaging to identify rice seed varieties [38]. Liu Weihua et al. integrated the advantages of the sparse features of the least absolute shrinkage and selection operator (LASSO) algorithm and the classification features of the logistic regression model (LRM). A hyperspectral rice seed purity identification method based on the LASSO logistic regression model (LLRM) was proposed. the recognition accuracy and average recognition accuracy of the LLRM were 91.67–100 and 98.47%, respectively. In addition, the identification accuracy of full-band LRM was 71.60–100%. However, the average identification accuracy was only 89.63% [39].

Xiangtao He et al. used near-infrared hyperspectral imaging (NIR-HSI) technology, combined with multiple data preprocessing methods and classification models, to identify the vigor of rice seeds. The experimental results showed that NIR-HSI technology has great potential in identifying rice seed viability and vigor [40]. Yanlin Wei et al. established a robust and accurate nondestructive soybean classification method through hyperspectral imaging and integrated machine learning algorithms. One of the integrated classifiers, the random subspace linear discriminant (RSLD) algorithm, was used to classify soybean seeds. Compared with linear discriminant (LD) and linear support vector machine (LSVM) methods, the results show that the RSLD algorithm is more stable and reliable. The RSLD method achieved the highest classification accuracy when classifying soybeans into 10, 15, 20 and 25 categories. When 155 features were used to classify 15 types of soybeans, the RSLD method achieved a classification accuracy of 99.2%, whereas the classification accuracy of the LD and LSVM methods was only 98.6 and 69.7% [41], respectively. Li Hao et al. proposed a soybean seed variety differentiation model based on hyperspectral imaging technology and one-dimensional convolutional neural network (1D CNN). A total of 3,600 hyperspectral images of soybean seeds (900 seeds per variety) were collected in the spectral range of 866.4–1,701.0 nm. Based on different numbers of sample sets, traditional machine learning models (k-nearest neighbor, support vector machine, partial least squares discriminant analysis) and a 1D CNN model were built, and the average accuracy of the training set was is higher than 96%, and the average accuracy of the test set is higher than 84% [42]. Zhu Susu et al. used near-infrared hyperspectral imaging to identify seven varieties of cottonseed. The scoring images formed by pixel principal component analysis (PCA) showed that there were differences between different varieties of cottonseed. Self-designed convolutional neural network (CNN) and residual network (ResNet) were used to build a classification model [43]. Hyperspectral imaging was used for peanut seed variety classification and mold detection by Wu Qingsong et al. The MF-LightGBM-SEL (Median Filter-Light Gradient Boosting Algorithm-Integrated Learning Model) model based on hyperspectral data achieved 98.63% and 98.03% prediction accuracies for data training and data testing, respectively, with a modeling time of only 0.37s [44] Sun Xia et al. captured the 1,000–2,030 nm band range in the hyperspectral images of four peanut kernels in a large black box, and based on the SPA-LDA (successive projection algorithm—linear discriminant analysis) model, the visual detection of peanut seed molds was successfully accomplished by using the 1,120 nm band and the thresholding method [45].

Hyperspectral image technology combines the strengths of traditional image technology and spectral technology. It allows for the selection of optimal bands for target recognition and facilitates the development of fast online detection or classification systems based on a few effective bands. However, hyperspectral images consist of numerous bands, resulting in large volumes of data. Therefore, the extraction of meaningful features from this extensive dataset and the application of data dimensionality reduction algorithms are crucial aspects of hyperspectral image research.

1.4 Summary and Outlook

The current global situation is relatively volatile, and factors such as population growth, limited land resources, and climate change have significantly impacted the agricultural sector. As a result, improving crop yields has become a pressing concern for countries worldwide. The presence of impurity seeds in crop seeds during sowing and grading operations can potentially affect harvest yields. Therefore, precise identification and detection of crop seed appearance are critical in helping farmers select high-quality seeds that are well-suited to specific environments, exhibit strong resistance to pests and diseases, and reduce the need for pesticides and fertilizers. Ensuring the quality and safety of agricultural products is key to enhancing crop yields. Currently, with the acceleration of urbanization and the global wave of digital transformation, agricultural production faces the challenges of transitioning to information technology, improving efficiency, and alleviating labor pressure.

Traditional crop seed appearance identification relies heavily on manual labor, relying on the naked eye to distinguish between good and bad seeds and perform manual sorting. This approach is time-consuming, costly in terms of labor, and prone to subjective errors. With the continuous advancements in science and technology, machine vision and hyperspectral technology have gradually become the mainstream methods for crop seed appearance recognition. They provide more efficient, accurate, and objective solutions for agricultural production.

Although these technologies have made significant progress, challenges persist for future development. On one hand, the integration of deep learning and artificial intelligence will further enhance the intelligence level of crop seed appearance recognition. The system will be able to automatically learn and optimize recognition models to improve accuracy and adaptability. On the other hand, the integrated application of multimodal sensing technologies, such as combining machine vision, hyperspectral, and X-ray technologies, can provide more comprehensive information for seed detection and enhance the reliability of the system.

However, in current artificial intelligence, deep learning machine vision neural networks heavily rely on large-scale labeled datasets and require a substantial number of labeled seed images for training. Obtaining high-quality labeled datasets poses a challenge. The design and selection of deep learning models and parameter tuning require significant time and computational resources. Constructing models for seed appearance recognition necessitates continuous optimization and algorithmic improvements to increase accuracy and robustness. Seeds exhibit diverse shapes, colors, textures, and other features, requiring the design of more complex and resilient models to handle the appearance characteristics of various seed types. Hyperspectral technology is sensitive to lighting conditions, and changes in lighting can impact the quality and spectral features of seed images. Therefore, attention must be given to controlling ambient light when collecting and analyzing seed images.

In conclusion, with the development of Internet of Things technology, future advancements in crop seed identification technology will focus on automation, intelligence, and integration. This will provide strong technical support for agricultural production, promote the widespread adoption of smart agriculture, ensure food security, and foster the sustainable development of agriculture.

References

1. Bates D, Morris J (2004) Application of machine vision systems in agriculture: A survey. Int J Agric Biol Eng 7(3):1–12
2. Patel D, Kumar A (2020) Spectral imaging for seed type classification: A spectral analysis approach. SensS Agric 15(1):34–49
3. Martin C, Davis P (2018) Machine vision systems for seed sorting: State of the art. Ind Crops Prod 112:749–758
4. Lee K, Park S (2022) Deep learning models for seed quality assessment: A review. Comput Electron Agric 195:106854
5. Gomez FR, Lee YS (2021) Convolutional neural networks for seed identification: A comprehensive analysis. Agric Inform 34(4):456–473
6. Nelson SO (2006) Electromagnetic and optical spectroscopy for quality assessment of seeds and kernels. J Agric Food Chem 54(23):8706–8713
7. Larson K, White J (2008) Deep learning approaches to challenges in seed phenotyping. Front Plant Sci 9:1162
8. Wu S, Sun D (2014) Computer vision technology for vegetable seed grading: A review. J Food Eng 123:165–172
9. Kim DH, Park JH (2013) Seed classification by machine learning and image analysis. Int J Precis Agric 14(1):62–77
10. Dana W, Ivo W (2008) Computer image analysis of seed shape and seed color for flax cultivar description. Comput Electron Agric 61(2):126–135
11. Yan X, Wang J, Liu S. et al (2011) Purity identification of maize seed based on color characteristics, Comput Comput Technol Agric IV: Springer, 620–628
12. Manickavasagan A, Sathya G, Jayas DS et al (2008) Wheat class identification using monochrome images. J Cereal Sci 47(3):518–527
13. Gunasekarans S, Cooper TM et al (1987) Image processing for stress crack in corn kernels. Trans ASAE 30(1):266–271
14. Neuman MR, Sapirstein HD, Shwedyk E, Bushuk W (1989) Wheat grain colour analysis by digital image processing II. Wheat class discrimination. J CeR Sci, 10:183–188
15. Liao K, Reid JJ, et al (1992) Corn kernel shape identification by machine vision using a neural network classifier. ASAE Peper. 7017
16. Chtioui Y, Panigrahi S, Backer LF (1999) Rough sets theory as a pattern classification tool for quality assessment of edible beans. Trans ASAE 42(4):1145–1152
17. Hecang Z, Qiaoli Z, Qing Z, et al. (2021) The design and experiment of peanut high-throughput automatic seed testing system based on machine learning. Acta Agric Scand, Section B—Soil Plant Science, 71(9):931–938

18. Balaji V, WJR, Pritam C (2018) DeepSort: deep convolutional networks for sorting haploid maize seeds. BMC Bioinform, 19(S9):289
19. Kurtulmuş F (2020) Identification of sunflower seeds with deep convolutional neural networks. J Food Meas Charact, 15(2):1–10
20. Shima J, Seyed-Hassan AM, JFV, et al (2021) Computer-vision classification of corn seed varieties using deep convolutional neural network. J Stored Prod Res, 92
21. Sidnal N, Patil UV, Patil P (2013) Grading and quality testing of foodgrains using neural network. Int J Res Eng Technol. 2(11):545–549
22. Zhao G, Quan L, Li H, Feng H, Li S, Zhang S, Liu R (2021) Real-time recognition system of soybean seed full-surface defects based on deep learning. Comput Electron Agric 187:106230
23. Song, Chang, Bo Peng, Huanyue Wang, Yuhong Zhou, Lei Sun, Xuesong Suo, **aofei Fan (2023) Maize seed appearance quality assessment based on improved Inception-ResNet. Front Plant Sci 14: 1249989
24. OuYang, Ai-Guo, Rong-jie Gao, Xu-dong Sun, Yuan-yuan Pan, **ao-ling Dong (2010) An automatic method for identifying different variety of rice seeds using machine vision technology. In 2010 Sixth International Conference on Natural Computation, 1, pp 84–88. IEEE
25. Bi C, Hu, Nan, Zou Y, Zhang S, Xu, Suzhen, Yu, Helong (2022) Development of deep learning methodology for maize seed variety recognition based on improved swin transformer. Agronomy 12(8):1843
26. Liu, Zhicheng, Long Wang, Zhiyuan Liu, Xufeng Wang, Can Hu, Jianfei **ng (2023) Detection of cotton seed damage based on improved YOLOv5. Processes 11(9): 2682
27. Xing X, Liu C, Han J, Feng Q, Lu, Qinglin, Feng Y (2023) Wheat-Seed variety recognition based on the GC_DRNet model. Agriculture 13(11):2056
28. Agelet LE, Hurburgh Jr CR (2014) Limitations and current applications of near infrared spectroscopy for single seed analysis. Talanta
29. Delwiche S (2003) Classification of scab-and other mold-damaged wheat kerels by near-infrared reflectance spectroscopy. Trans-Am Soc Agric Eng 46(3):731–738
30. Vermeulen P, Pierna JF, Buhigiro T, Herman JL, Lecler B, Mouteau A, Dardenne P (2007) Barley varieties discriminated by the near infrared hyperspectral imaging technique. In 13th International Conference on Near Infrared Spectroscopy, umea, Sweden (pp. 15–21)
31. Yang, Huan, Cheng Wang, Han Zhang, Ya'nan Zhou, and Bin Luo (2023) Recognition of maize seed varieties based on hyperspectral imaging technology and integrated learning algorithms. Peer J Computer Science 9: e1354
32. Zhu, Shaolong, Maoni Chao, **yu Zhang, **njuan Xu, Puwen Song, **long Zhang, Zhongwen Huang. (2019) Identification of soybean seed varieties based on hyperspectral imaging technology. Sensors 19(23): 5225
33. Jing B, Zhang C, Jia L, Lu, Qizhe Tang, Gao GZ, Qi H (2022) Identification of rice seed varieties based on near-infrared hyperspectral imaging technology combined with deep learning. ACS Omega 7(6):4735–4749
34. Zhang, Han, Qiling Hou, Bin Luo, Keling Tu, Chang** Zhao, Qun Sun (2022) Detection of seed purity of hybrid wheat using reflectance and transmittance hyperspectral imaging technology. Front Plant Sci 13: 1015891
35. Ambrose, Ashababebwa, Lalit Mohan Kandpal, Moon S. Kim, Wang-Hee Lee, Byoung-Kwan Cho (2016) High speed measurement of corn seed viability using hyperspectral imaging. Infrared Phys & Technol 75 (2016): 173–179
36. Sun J, Zhang L, Zhou X, Yao K, Tian Y, Nirere A (2021) A method of information fusion for identification of rice seed varieties based on hyperspectral imaging technology. J Food Process Eng 44(9):e13797
37. Zhu, Qibing, Zhaoli Feng, Min Huang, and **ao Zhu (2012) Maize seed classification based on image entropy using hyperspectral imaging technology. Trans Chin Soc Agric Eng, 28(23): 271–276
38. Qiu Z, Chen J, Zhao Y, Zhu S, He Y, Zhang C (2018) Variety identification of single rice seed using hyperspectral imaging combined with convolutional neural network. Appl Sci 8(2):212

39. Liu W, Zeng S, Wu, Guiju, Li H, Chen F (2021) Rice seed purity identification technology using hyperspectral image with LASSO logistic regression model. Sensors 21(13):4384
40. He X, Feng X, Sun D, Liu F, Bao Y, He Y (2019) Rapid and nondestructive measurement of rice seed vitality of different years using near-infrared hyperspectral imaging. Molecules 24(12):2227
41. Wei, Yanlin, **aofeng Li, **n Pan, Lei Li (2020) Nondestructive classification of soybean seed varieties by hyperspectral imaging and ensemble machine learning algorithms. Sensors 20(23): 6980
42. Li H, Zhang L, Sun H, Rao Z, Ji H (2021) Identification of soybean varieties based on hyperspectral imaging technology and one-dimensional convolutional neural network. J Food Process Eng 44(8):e13767
43. Zhu S, Zhou L, Gao P, Bao Y, He Y, Feng L (2019) Near-infrared hyperspectral imaging combined with deep learning to identify cotton seed varieties. Molecules 24(18):3268
44. Wu Q, Xu, Lijia, Zou Z, Wang J, Zeng Q, Wang Q, Zhen J, Wang Y, Zhao Y, Zhou M (2022) Rapid nondestructive detection of peanut varieties and peanut mildew based on hyperspectral imaging and stacked machine learning models. Front Plant Sci 13:1047479
45. Sun X, Liu J, Sun J, Zhang H, Guo Y, Zhao W, **a, L, Wang, B. (2021) Visual detection of moldy peanut kernels based on the combination of hyperspectral imaging technology and chemometrics. J Food Process Eng 44(11):e13864

Chapter 2
A Review of Corn Growth Status Sensing Methods

Xuan Hu, Wenyi Sheng, Zhao Zhang, Ruicheng Qiu, and Man Zhang

Abstract Corn is one of the most important food crops in the world, widely cultivated across the globe, and extensively used in food, feed, and industrial materials. With the growth of the global population and increasing food demand, improving the yield and quality of corn has become a significant challenge in agricultural production. In this context, accurately sensing the growth status of corn is crucial for achieving precision agriculture, enhancing yield and quality, and avoiding soil pollution caused by excessive fertilization. This review first introduces the evaluation indicators of corn growth status, such as leaf area index and nitrogen content index, and then classifies the existing sensing methods into manual measurement and remote sensing measurement. It further discusses data processing methods, including preprocessing, feature extraction, and modeling techniques. Finally, it provides an outlook on future developments, aiming to offer a reference for the intelligent and precise production of agriculture.

Keywords Corn growth · Growth evaluation index · Non-Destructive inspection · Remote sensing technique · Spectral analysis

2.1 Introduction

Crops are fundamental to human survival and development, underscoring their paramount importance [1]. Based on the annual grain import volumes, China is gradually becoming the world's largest grain importer [2]. As one of the most critical food crops globally, corn is extensively planted and produced in China and holds a significant position in global agricultural production. Corn production directly affects national food security and farmers' economic interests [3]. Besides being a

X. Hu · W. Sheng (✉) · Z. Zhang · R. Qiu · M. Zhang
College of Information and Electrical Engineering, China Agricultural University, Beijing 100083, China
e-mail: wenyi.sheng@cau.edu.cn

Key Laboratory of Smart Agriculture System Integration, Ministry of Education, China Agricultural University, Beijing 100083, China

Y. Yun et al. (eds.), *Advanced Sensing and Robotics Technologies in Smart Agriculture*, Smart Agriculture 9, https://doi.org/10.1007/978-981-97-6441-9_2

crucial food and feed source, corn is widely used in industrial production, such as bioenergy and plastic substitutes.

With population growth and economic development, the demand for corn yield and quality continues to increase, making the sensing and monitoring of corn growth status particularly important. Traditional corn production management often relies on experience and intuition, lacking scientific data support. However, with the advancement of modern agricultural technology, sensing the growth status of corn has become key to improving its yield and quality. Early awareness of crop growth conditions during the plant growth period can be more critical than accurately estimating planting area and total yield, especially for potential large-scale food shortages or surpluses. Timely acquisition of crop growth information is crucial.

By sensing the growth status of corn, farmers can promptly understand crop conditions and formulate more scientific planting strategies, thereby enhancing corn yield and quality. Additionally, sensing the growth status of corn provides data support for agricultural production, facilitating precision agriculture management and contributing to the sustainable development of agriculture. Real-time monitoring and analysis of corn growth can help identify and solve problems early, reducing resource waste and environmental pollution. Overuse of fertilizers is a primary cause of soil acidification [4]. Soil is the foundation of agricultural production; in acidified soils, the activity of heavy metal ions significantly increases, leading to heavy metal contamination of crops. Excessive fertilizer use also causes non-point source pollution of soil, severely impacting the quality and safety of agricultural products and daily life [5]. Furthermore, sensing the growth status of corn can support agricultural production with data, aiding in precision agriculture management and improving agricultural production efficiency and quality.

2.2 Indicators of Corn Growth Status

The evaluation of corn growth status primarily relies on a series of indicators, including plant height, leaf area index, chlorophyll content index, and vegetation index. These indicators can reflect the growth and physiological status of corn, playing a crucial role in predicting corn yield and formulating planting strategies.

2.2.1 Plant Height

Plant height refers to the vertical distance from the ground to the top of the highest leaf of a corn plant. It is one of the important indicators of plant growth status. Plant height directly reflects the growth of the corn plant; changes in plant height can indicate the growth rate and potential of the corn. Additionally, plant height can indirectly reflect factors such as soil fertility and environmental conditions affecting plant growth. Under normal growth conditions, corn plants typically have a height ranging from 1.2 to 2.4 m from emergence to harvest. Abnormal plant height, whether too tall or too short, may indicate growth issues. For instance, short plant height may be caused by pests, nutrient deficiency, water shortage, or poor soil conditions, while excessively tall plants may be more prone to lodging.

Plant height is closely related to yield. By analyzing plant height data, crop yield can be predicted, providing a basis for harvest and storage decisions. In the breeding process, plant height is also an important selection criterion. By selecting varieties with moderate height and strong lodging resistance, crop yield and stability can be improved.

2.2.2 Leaf Area Index

Leaf Area Index (LAI) refers to the total leaf area of plants per unit land area and is used to describe the capacity of a plant canopy to intercept light energy. It is an important indicator for evaluating the photosynthetic efficiency and light energy utilization rate of corn [6]. LAI is calculated by measuring the total area of corn leaves [7]. This index reveals the corn plants' ability to absorb light energy and their growth potential [8].

During the growth period of corn, changes in LAI are closely related to the intensity of photosynthesis, which helps farmers make timely decisions on fertilization, pest control, and field management. Both excessively high and low LAI can affect the yield and quality of corn. A low LAI may indicate insufficient leaf area, impacting photosynthetic efficiency and yield, while a high LAI may cause leaves to shade each other, reducing light use efficiency.

Monitoring changes in LAI allows for understanding the photosynthetic capacity and growth status of corn. By measuring LAI under different fertilization levels, optimal fertilization amounts and timings can be determined, thereby improving fertilizer utilization and crop yield [9]. Similarly, measuring LAI at different planting densities helps determine the optimal planting density, making full use of light energy to enhance crop yield and quality.

2.2.3 Chlorophyll Content Index

The Chlorophyll Content Index reflects the chlorophyll content in corn leaves, with chlorophyll being a crucial pigment for photosynthesis in corn. Its content directly influences the photosynthetic efficiency and growth rate of corn. The Chlorophyll Content Index can be obtained by measuring the chlorophyll content in corn leaves. Healthy corn plants typically have a higher Chlorophyll Content Index. Changes in the Chlorophyll Content Index can reflect the growth status and physiological changes of corn. An abnormal Chlorophyll Content Index may indicate that corn leaves are affected by pests, nutrient deficiency, or environmental factors, leading to reduced photosynthetic efficiency.

2.2.4 Nitrogen Content Index

Nitrogen is one of the essential nutrients for corn growth [10]. The Nitrogen Content Index is an indicator of assessing the nitrogen nutrition status of corn plants. It reflects the nitrogen nutrition status of corn plants by measuring the nitrogen content in leaves or stems, playing a significant role in guiding fertilization and improving nitrogen fertilizer utilization efficiency. A low Nitrogen Content Index may indicate insufficient nitrogen nutrition in corn plants, leading to growth limitations and reduced yield, while an excessively high Nitrogen Content Index may result in nutrient excess and environmental pollution.

2.2.5 Other Indicators

Other indicators include stem diameter, root growth status, etc., which can also reflect the growth status of corn from different perspectives. The thickness of the stem can reflect the robustness, lodging resistance, and yield potential of corn. Variations in stem thickness within a limited range can reflect differences in the growth status and yield potential of corn. Corn plants with moderate and uniform stem thickness typically have better growth status and yield potential. By monitoring stem growth, management measures such as increasing support and adjusting planting density can be timely adjusted.

Roots are the main organs for corn to absorb water and nutrients, and their growth status can reflect the adaptability and yield potential of corn. Monitoring root growth by excavating roots or using non-destructive testing techniques (such as CT scans) can provide scientific basis for management measures such as fertilization and irrigation.

Leaves are the intuitive expression of corn leaf color and the main organs for corn photosynthesis. Their health directly affects corn growth and yield. Leaf color can reflect the health and nutritional status of corn plants, with healthy corn plants typically exhibiting deep green leaves. By monitoring leaf color, shape, and pest/disease conditions, leaf problems can be promptly discovered and addressed to ensure the healthy growth of corn. Yellowing or wilting of corn leaves may indicate that plants are affected by adverse factors such as excessive nitrogen fertilizer, insufficient water, improper soil pH, etc.

2.3 Measurement Methods

Currently, methods for sensing the growth status of corn mainly include manual measurement and remote sensing measurement.

2.3.1 Manual Measurement

Undoubtedly, manual observation is the oldest method for monitoring corn growth and is currently the most widely used method by agricultural producers. Observers judge the growth of corn by observing its geometric dimensions, shape, color, and other external characteristics, such as water deficiency, nutrient deficiency, pest and disease damage, etc. According to plant nutrition principles, leaf color is the most sensitive indicator of nitrogen nutrition status, with leaf color positively correlated with nitrogen content in leaves [11]. When plants are deficient in nitrogen, protein synthesis decreases, enzyme and chlorophyll content decrease, cell division slows down, and leaves turn yellow, resulting in premature maturation and low yield. Conversely, when plants have excess nitrogen, carbon metabolism is uncoordinated, protein synthesis increases, carbohydrates are consumed in large quantities, and cellulose decreases, leading to elongation, reduced resistance, susceptibility to disease and lodging, and reduced yield. People have long mastered the method of judging plant nutrition status from their leaves based on practice. People have accumulated a lot of knowledge about judging plant growth conditions. For example, leaf color can indicate moisture conditions, and spots or abnormalities on leaves and stems can indicate pest and disease damage. Although this method is simple and direct, it is time-consuming and labor-intensive, and the measurement accuracy is greatly influenced by human factors. The subjectivity of the observation results makes it unsuitable for large-scale monitoring.

With the development of science and technology, some intelligent and informatized testing techniques have been developed, and plant growth diagnosis has gradually developed towards plant nutrition diagnosis based on crop growth conditions. For example, to facilitate rapid diagnosis of chlorophyll nutrition status of field-grown crops, Japan has developed and produced the handheld SPAD-502 chlorophyll meter (Fig. 2.1), which is based on the strong absorption of red light and low absorption of far-red light [12]. It has the advantages of simple operation, rapid data acquisition, no consumables required, and no negative effects on plants and the environment. The SPAD value estimates the amount of chlorophyll per unit weight of leaves. In the field, the chlorophyll status of crops can be estimated by testing the leaf greenness of plants, and it has been applied and promoted in various crops such as cotton, rice, wheat, corn, and barley [13–17]. However, since SPAD readings are greatly affected by light radiation intensity [18], and this method requires contact measurement and testing of multiple plants to obtain their average as the measurement result, it has the disadvantages of large workload. Therefore, finding a more convenient, rapid, and non-contact instrument-based method to determine crop growth conditions has become a focus of research.

The GreenSeeker (Fig. 2.2), developed by NTech Industries in the United States, utilizes optical principles to monitor crop growth. This device employs an active light source and non-contact method to measure the spectral reflectance of crops. By calculating vegetation indices such as NDVI, DVI, RVI, and others, it analyzes crop growth status and potential yield, among other factors.

Lukina et al. proposed the Nitrogen Fertilization Optimization Algorithm (NFOA), the core of which lies in predicting potential yield and the crop's nitrogen uptake based on the Normalized Difference Vegetation Index (NDVI) of field crops [19]. By real-time acquisition of wheat canopy reflectance spectra using optical sensors, the device calculates potential yield per unit area and nitrogen application rates. Variable rate fertilizers are then applied directly in the field according to different fertilization levels. In terms of detecting chlorophyll content in large plant

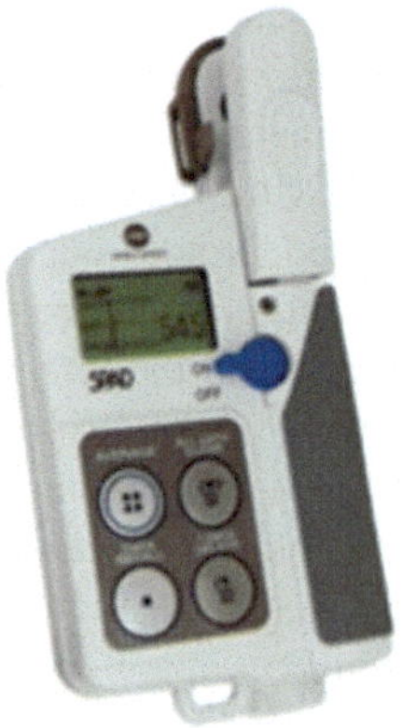

Fig. 2.1 SPAD-502Plus handheld chlorophyll meter produced by Konica Minolta, Japan

Fig. 2.2 GreenSeeker field measurement diagram developed by NTech Industries, USA

leaf populations, the CM1000 Chlorophyll Meter developed by Spectrum Technologies in the United States utilizes "calibration measurement" technology to quickly measure reflectance at wavelengths ranging from 700 to 840 nm, detecting important physiological indicators during critical growth stages of crops [20, 21]. The CM1000 Chlorophyll Meter uses laser positioning to mark the boundaries of the sampled leaves or agricultural areas (Fig. 2.3). It has a built-in ambient light sensor to correct interference factors under different lighting conditions, allowing for the detection of nutrient deficiencies in crops before growth issues become visually apparent. However, the instrument's built-in battery has limited capacity and requires recharging, making it unsuitable for prolonged field use without access to power sources.

Kitić et al. proposed a low-cost, portable active multispectral optical device for precise plant stress detection and field mapping, called the Plant-O-Meter (Fig. 2.4). This device features an integrated multispectral source, including light sources at four wavelengths (850, 630, 535, and 465 nm), capable of simultaneously illuminating the entire plant [22].

In China, Zhao et al. developed a Normalized Difference Vegetation Index (NDVI) meter using sunlight as the light source [23]. It utilizes four photodetectors with

Fig. 2.3 CM1000 chlorophyll detector developed by Spectrum Technologies, USA

Fig. 2.4 Plant-O-Meter active light source Multispectral sensor and external GPS antenna connected to smartphone via Bluetooth

special spectral response characteristics to detect the reflected light from vegetation at two characteristic wavelengths in the near-infrared and red light regions. After analog-to-digital conversion, the NDVI is calculated, and using data from the NDVI meter, they successfully monitored wheat plant leaf area and chlorophyll density.

At the National Engineering Research Center for Information Technology in Agriculture, Nanjing Agricultural University, Ni et al. developed a portable crop growth monitoring and diagnostic device based on a multispectral sensor [24]. The device is designed with a passive light source, and the optical channel underneath measures the spectral reflectance intensity of crop leaf canopies at wavelengths of 710 nm and 820 nm, while the optical channel above measures solar irradiance (Fig. 2.5). Internally, a microcontroller is used to calculate reflectance and compute various vegetation indices displayed on the LCD screen. Experiments conducted using the multispectral sensor on crops such as rice demonstrated that the instrument can non-destructively estimate important growth indicators such as chlorophyll content in rice.

Ding et al. proposed two design criteria for spectrometers based on active light sources (Fig. 2.6), using 730 nm and 810 nm as the central wavelengths for detection bands [25]. They tested the Normalized Difference Vegetation Index (NDVI) and Ratio Vegetation Index (RVI) of wheat fields under two levels of nitrogen application rates (NARLs).

Fig. 2.5 Nitrogen measurement sensor of Nanjing Agricultural University

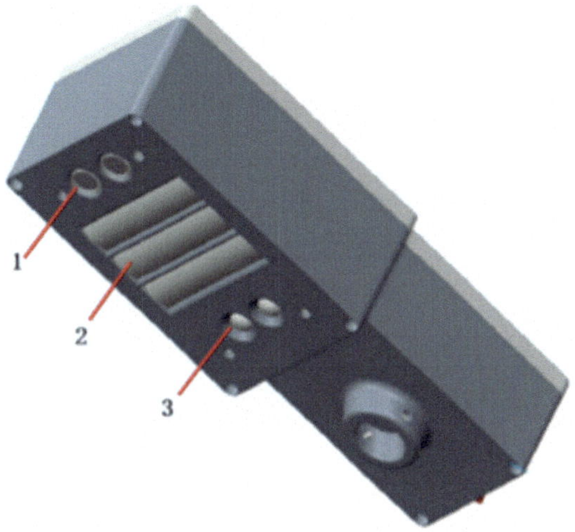

Fig. 2.6 Ding independently developed dual-band reflection spectrometer (1) ultrasonic ranging sensor, (2) narrow-band active light source, (3) reflective light detector

2.3.2 Remote Sensing Measurement

The rapidly advancing remote sensing technology provides a new scientific means for the dynamic monitoring of crop growth and yield estimation by offering macroscopic, dynamic, and rapid observations of the Earth (Fig. 2.7). It can be equipped with sensors such as multispectral cameras, LiDAR, etc., to obtain multidimensional information about the growth status of corn. Through remote sensing technology, extensive information about the growth status of corn can be obtained. Remote sensing data has advantages such as wide coverage, large information content, and strong real-time capabilities, but its interpretation is complex and is significantly influenced by factors such as cloud cover and atmosphere.

Crop growth remote sensing monitoring is based on the spectral theory of green plants. According to the reflection characteristics of green plants in the spectrum, there is a strong absorption band in the visible light part and a strong reflection peak in the near-infrared part, which reflects the growth information of crops and thereby judges the growth status of crops for monitoring purposes [27]. Currently, most studies use land satellite remote sensing data (Landsat TM) and higher-resolution meteorological remote sensing data (NOAA AVHRR), while also developing crop growth monitoring using hyperspectral satellite remote sensing and radar remote sensing. After processing the captured images, crop growth remote sensing indicators such as Leaf Area Index (LAI), Chlorophyll Content (CHL), Normalized Difference Vegetation Index (NDVI), etc., are extracted to assess crop growth status and detect nutrient deficiencies and pest situations early, thereby taking proactive and effective measures. Various stresses such as nitrogen deficiency and drought can alter the light reflection characteristics of plant leaves. The main factors affecting light absorption and reflection by leaves are chlorophyll, protein, water, and carbon-containing compounds, with chlorophyll content having the greatest impact. Remote

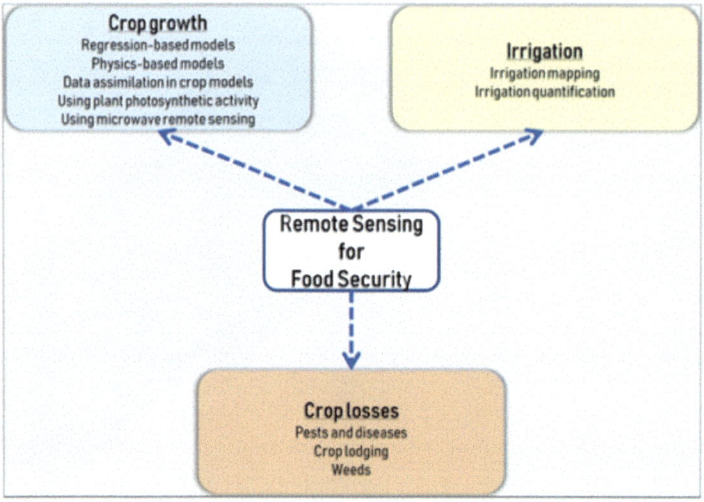

Fig. 2.7 Karthikeyan summarized the application of remote sensing in agricultural management [26]

sensing technology is utilized to monitor crop growth status by monitoring the reflection and absorption of light by crop canopies. Monitoring with wide-band and multi-band spectrometers has shown that vegetation indices generated from combinations of near-infrared (around 800 nm), red light (around 680 nm), and green light (around 560 nm) wavelengths can be used to monitor chlorophyll status in crops (such as rice, cotton, and maize). Additionally, remote sensing imaging maps of wheat growth obtained through hyperspectral imaging spectrometers have been widely promoted and applied for yield estimation and fertilization based on vegetation indices (Fig. 2.8). Since the development of remote sensing technology, countries around the world have actively explored its use in monitoring plant growth.

Guan et al. utilized two types of small unmanned aerial vehicles, the DJI Phantom 4 and DJI Phantom 4 Pro, each equipped with a compact multispectral sensor, to monitor designated areas and obtain NDVI values [29]. Song et al. [30] used aerial image data obtained from the Pushbroom Hyperspectral Imager (PHI) during the jointing, grain filling, and milk ripening stages of winter wheat as the basis to acquire the growth status and leaf area index of winter wheat, analyzing the growth situation of winter wheat in the study area [30]. The Remote Sensing Laboratory of the U.S. Department of Agriculture's Agricultural Research Service has been using filter-based multi-band cameras to capture images of target areas since 1982, synthesizing them into false-color images for interpretation and assessment of agricultural production issues such as pests and diseases, soil salinity, soil nutrient status, and water pollution [31]. With the development of remote sensing technology, some high-resolution aerial photographs have gradually been applied to agricultural production management. High-resolution aerial images and satellite images obtained using airborne visible-infrared imaging spectrometers (AVIRIS), compact airborne spectral imaging

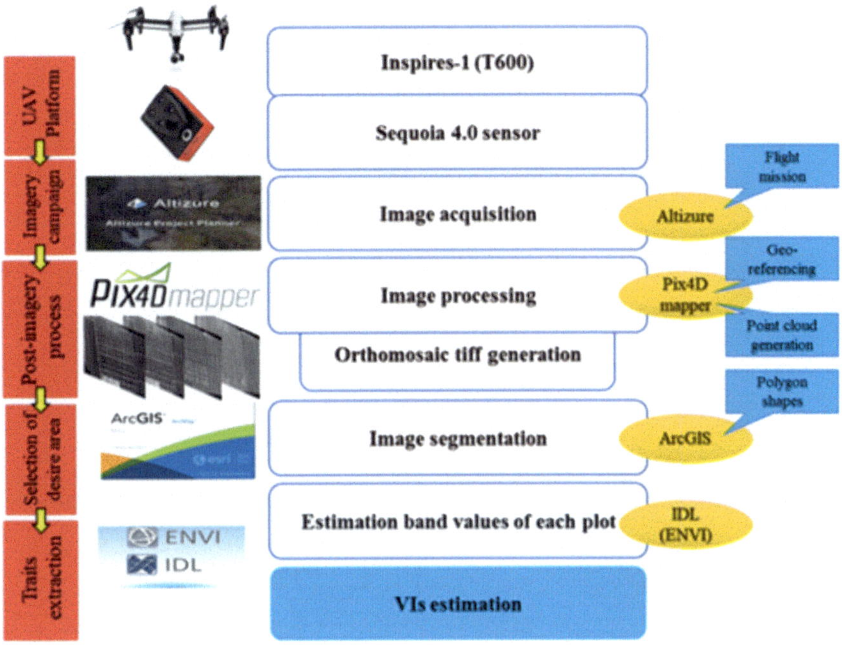

Fig. 2.8 The workflow diagram of a multi-spectral feature extraction method based on UAV proposed by [28]

(CASI), and other high-resolution aerial imaging spectrometers are also increasingly used for large-scale monitoring of crop growth conditions (Figs. 2.9, 2.10). Zhang et al. used unmanned aerial vehicles to collect crop images and combined them with deep learning algorithms to establish a rapid growth prediction model, achieving rapid monitoring of crop growth [32]. Liu et al. [33] obtained remote sensing data from UAV aerial images and proposed a new method for monitoring the growth of rapeseed based on multi-temporal mixed spectra. Maimaitijiang et al. evaluated the potential of using satellite/UAV data fusion and machine learning to combine canopy spectral information with canopy structure features for crop monitoring [34].

2.4 Data Processing

Currently, the collection of corn growth information both domestically and internationally primarily involves the acquisition of apparent information (such as visible physical data like corn plant height, stem diameter, etc.) and intrinsic information (such as leaf water content, chlorophyll content, nitrogen content, etc., obtained through external methods). Spectral technology, as a non-contact, non-destructive analytical method, has been widely applied in the process of sensing corn growth

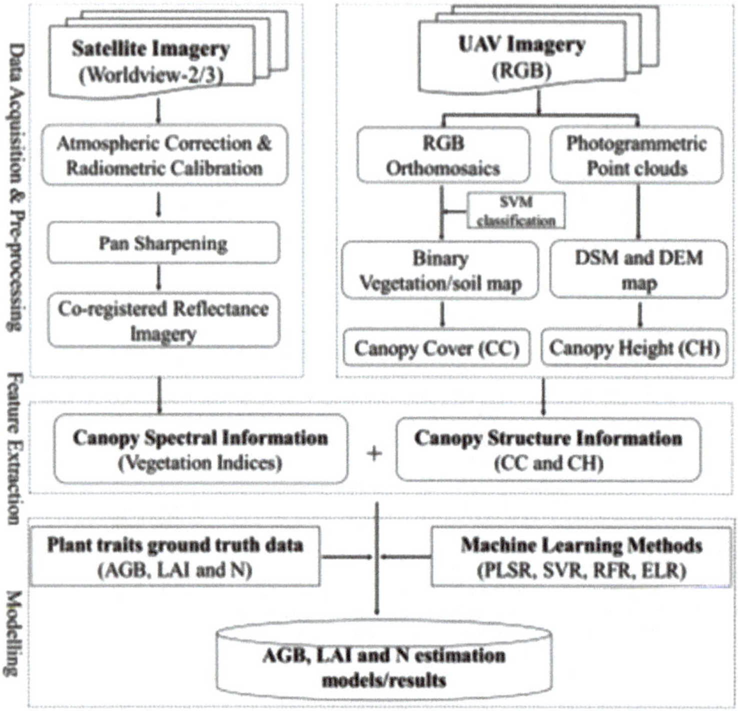

Fig. 2.9 The workflow flow chart of remote sensing image preprocessing, canopy spectral and structural feature extraction, model training and testing proposed by Maimaitijiang et al.

Fig. 2.10 The UAV (**a**), the multispectral camera (**b**) and the schematic diagram of the multispectral camera (**c**) used by [35]

status, as elaborated in the previous chapters [36]. Spectral data processing is a crucial step in spectral analysis technology, and its results directly affect the accuracy and reliability of spectral analysis. With the rapid development of computer technology and artificial intelligence, spectral data processing methods have also been continuously developed and improved.

Corn spectral data are obtained through spectral measurements of the light response of corn leaves, stems, and other tissues at different wavelengths using spectrometers. This data is rich in information, has high resolution, and can be measured quickly. Such data includes extensive information on corn growth, such as chlorophyll content, water content, and nitrogen content. However, due to the high dimensionality, redundancy, and complexity of spectral data, direct analysis and interpretation can be challenging. Therefore, effective processing and analysis of corn spectral data is essential [37].

2.4.1 Preprocessing

Spectral data preprocessing is the first step in spectral data processing. Typically, raw spectral data is obtained through spectral measurements of corn leaves, stems, and other tissues using spectroscopic instruments. During data collection, it is crucial to avoid interference from environmental noise and instrument errors. The purpose of preprocessing is to eliminate noise and background interference in the spectral data, thereby improving its quality. The following are some common spectral data preprocessing methods:

Noise is one of the inevitable interfering factors in spectral data. During the acquisition of corn spectral data, noise often appears due to instrument errors, environmental interference, and other factors. The presence of noise affects the accuracy and reliability of the data. Therefore, appropriate noise removal methods need to be employed during the preprocessing stage, with smoothing being the most commonly used method for eliminating spectral noise. Various mathematical methods for smoothing exist, such as Fourier Transform (FFT), Singular Value Decomposition (SVD), convolution smoothing methods, Kalman filtering, and others [38].

Baseline correction is an important step in eliminating background interference in spectral data. Baseline shift is a common phenomenon in spectral data that affects the accuracy and comparability of the data. The purpose of baseline correction is to eliminate baseline shift and make the spectral data more accurate. Common baseline correction methods include linear fitting, polynomial fitting, and others. Chen et al. used the BRUKER EQUINOX55 Fourier near-infrared spectrometer to measure the near-infrared absorption spectra of isobutane gas at different concentrations, finding that baseline correction made the experimental results more accurate [39].

To eliminate the interference of noise and baseline in measured spectral signals, Fang et al. proposed a new method based on wavelet transform for simultaneously removing both and selecting parameters [40]. This method achieves baseline correction by zeroing out the low-frequency wavelet coefficients in the wavelet domain of the spectral signal and removes noise by thresholding the high-frequency wavelet coefficients. Experiments show that this method is effective in eliminating noise and baseline in measured spectra.

2.4.2 Feature Extraction

Feature extraction is one of the key steps in spectral data processing, aiming to extract feature information related to the target analyte from the spectral data [41]. This feature information can be used for subsequent classification, identification, or quantitative analysis.

Spectral enhancement is a method used to highlight specific regions or features of the spectral curve. Common spectral enhancement methods include derivatives, wavelet transform, and others. By calculating the first or second derivatives of the spectral data, information such as peaks, valleys, or inflection points of the spectral curve can be highlighted, improving the resolution and sensitivity of the spectral data. Zhang et al. proposed a method for visible-near-infrared spectral analysis of phosphorus composition in fertilizers using a non-integer order Savitzky-Golay derivative function (NISGDF) [42]. The formula for NISGDF was developed and presented in matrix form.

Feature selection is the process of selecting the subset of features most relevant to the target analyte from the spectral data. Common feature selection methods include Principal Component Analysis (PCA), Independent Component Analysis (ICA), and others. These methods can identify the main variation patterns and independent components in the spectral data, providing useful feature information for subsequent classification or identification. Principal Component Analysis (PCA) is a dimensionality reduction technique that aims to transform the original high-dimensional data into a lower-dimensional feature space through linear transformation, thereby retaining the most important information in the data and reducing its complexity. In corn spectral data modeling, PCA can be used to reduce the dimensionality of the spectral data and extract the most important features. By retaining the most significant principal components, noise and redundant information in the data can be reduced, improving the efficiency and accuracy of modeling. Zhai et al. conducted qualitative Vis–NIR spectral detection of cadmium stress in lettuce leaves [43]. They preprocessed the collected spectral information using wavelet transform, first-order derivative, and second-order derivative methods, and then reduced the data dimensionality using Principal Component Analysis.

Peak identification is an important aspect of spectral data analysis, aiming to identify peaks in the spectral data and determine their corresponding substances. Common peak identification methods include Gaussian fitting, Lorentz fitting, and others. These methods can accurately determine the parameters such as the position, height, and width of the peaks.

2.4.3 Modeling

Data modeling is the process of using extracted features to build models that predict the growth status of corn. By utilizing the extracted feature information to classify or identify spectral data, rapid and accurate detection of samples or components can be achieved. Common modeling methods in data processing include statistical models and machine learning models. These methods can establish predictive models based on historical and current data to predict and analyze the growth status of corn.

Linear Regression (LR) is a statistical model used to establish a linear relationship between the dependent variable (target variable) and one or more independent variables (feature variables). It describes the relationship between the independent and dependent variables by fitting a straight line (or plane, hyperplane). In corn spectral data modeling, linear regression can be used to predict the linear relationship between corn growth status and spectral features. By fitting a linear model between spectral data and actual observations, simple and intuitive predictions can be made. Lu et al. utilized a low-cost drone system to capture images of wheat during critical growth stages at an altitude of 30 m in Rugao, eastern China, and processed the data using the Stepwise Multiple Linear Regression (SMLR) method [44].

Support Vector Machine (SVM) is a supervised learning algorithm that aims to find an optimal hyperplane that maximally separates samples of different classes. By finding this optimal hyperplane in the feature space, SVM can classify new samples or make regression predictions. In corn spectral data analysis, SVM can be used for classification tasks, such as predicting different growth stages or pest status based on corn spectral data. By training the SVM model, corn spectral data can be mapped into a high-dimensional space to achieve accurate predictions of corn growth status. Sun et al. conducted pot experiments to measure the chlorophyll content of corn leaves at different treatment times using spectral methods [45]. They used the Least Squares Support Vector Machine (LSSVM) algorithm to estimate the chlorophyll content of corn leaves. Xiang et al. used multispectral remote sensing by drones to monitor soybeans in the northwest region, selecting five vegetation indices related to soybean leaf area index, and constructed an estimation model using SVM, which was then validated [46].

Random Forest (RF) is an ensemble learning method that constructs multiple decision trees and combines their predictions for classification or regression. Each decision tree is built based on different random samples and features, and the final prediction result is determined by the voting or averaging of all decision trees. In corn spectral data modeling, RF can be used for classification and regression tasks. By training multiple decision trees and combining their predictions, the robustness and generalization ability of the model can be improved, leading to more accurate predictions of corn growth status or pest conditions. Song et al. monitored wheat during different growth stages using drone remote sensing and found that the RF regression model performed best during the jointing, booting, and heading stages, with R^2 values of 0.67, 0.43, and 0.70, respectively [47]. Liu et al. used multispectral images from drones to study the LAI values of cotton leaves and found that the RF regression model was the best model [48].

Artificial Neural Networks (ANNs) are networks composed of multiple neurons (nodes) that simulate the working principles of the human nervous system. By adjusting the connection weights and activation functions between neurons, ANNs can learn complex nonlinear mapping relationships between inputs and outputs. In corn spectral data modeling, ANNs can be used for classification and regression tasks. By training multi-layer neural networks and adjusting network structures and parameters, accurate predictions of corn growth status can be achieved. ANNs have strong expressive ability and adaptability when dealing with large-scale data and complex features. Zhao et al. used multispectral cameras on drones to collect data, selected eight vegetation indices, and used three methods—linear regression, random forest, and artificial neural network—to model different indices, finding that the prediction accuracy of the ANN model was significantly better than that of the linear regression and random forest models [49]. Xie et al. used Sentinel-2 data to retrieve biophysical parameters such as the leaf area index (LAI), leaf chlorophyll content (LCC), and canopy chlorophyll content of winter wheat using the ANN method [50].

In addition to the methods mentioned above, there are many other methods or multi-variate models. Yan and Xu et al. used hyperspectral images of each leaf for three-dimensional analysis (Fig. 2.11), employing Convolutional Neural Networks (CNNs) for identification and comparing them with traditional machine learning methods [51]. For the extracted spectra, 1D CNN showed good classification performance with an accuracy of up to 98%. For RGB images, 2D CNN showed better classification performance. For hyperspectral images, 3D CNN performed moderately, better than 2D CNN.

Luo et al. utilized several vegetation indices and spectral parameters from the Spectral Continuum Removal Method (SPCRM) to invert the Leaf Area Index

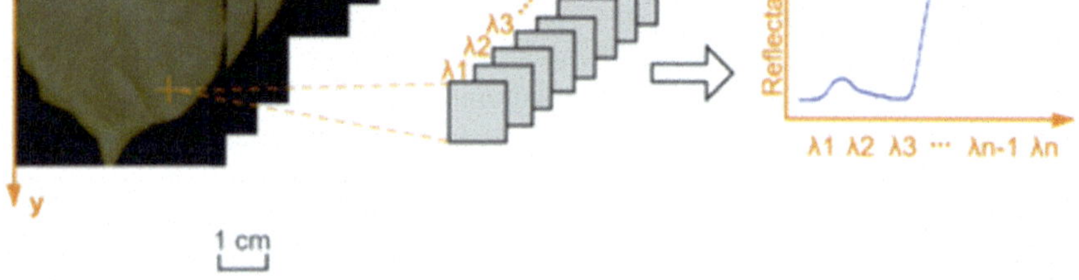

Fig. 2.11 Three-dimensional data cube diagram of hyperspectral image of cotton leaves in Yan and Xu

(LAI) of potatoes [52]. The coefficients of determination for the retrieval models of total area, left side area, right side area, and depth area ratio were all greater than 0.801, indicating that SPCRM has the potential for accurate retrieval of potato LAI. Zheng et al. improved the estimation of rice Above Ground Biomass (AGB) by combining texture and spectral analysis of UAV images [53]. Simple regression and stepwise multiple linear regression models were established between the two-year experimental data and the image parameters of four different types of feature sets. The results showed that the combination of texture information with spectral information significantly improved the accuracy of rice biomass estimation compared to using spectral information alone.

2.5 Summary and Outlook

Perceiving the growth status of corn is crucial for improving its yield and quality. Currently, methods such as manual measurements and remote sensing measurements are widely used to perceive the growth status of corn. With the continuous development of modern agricultural technology, methods for perceiving the growth status of corn are also constantly advancing. In the future, as artificial intelligence, the Internet of Things (IoT), and other technologies continue to integrate, the perception of corn growth status will become more intelligent and precise. Real-time monitoring and analysis of corn growth status can help detect and resolve issues promptly, thereby enhancing the yield and quality of corn. Additionally, with the application of big data and cloud computing technologies, the processing and analysis of corn growth status data will become more efficient and convenient. This will provide more accurate data support for agricultural production, driving the progress of agricultural modernization forward continuously.

References

1. Schmitz N, Yu N, et al. (2016) Development of methods to improve soybean yield estimation and predict plant maturity with an unmanned aerial vehicle based platform. Remote Sens Environ: Interdiscip J. https://doi.org/10.1016/j.rse.2016.10.005
2. Wei Y (2014) China is becoming the world's largest grain importer. Import Export Manag (1):4. https://doi.org/10.3969/j.issn.1673-6532.2014.01.009
3. Schirrmann M, Giebel A, et al. (2016) Monitoring agronomic parameters of winter wheat crops with Low-Cost UAV Imagery. Remote Sensing, 8(9). https://doi.org/10.3390/rs8090706
4. Shi Y (2002) Modern agriculture. China Agric Sci Bull 4:7–10. https://doi.org/10.3969/j.issn.1008-0864.2002.06.002
5. Wang W, Yang Z, et al. (2015) Overview of research on soil acidification in China. J Anhui Agric Sci 43(8):3. https://doi.org/10.3969/j.issn.0517-6611.2015.08.022
6. Chen N, Schlaepfer DR, et al. (2023) Evapotranspiration partitioning using a Process-Based model over a Rainfed Maize Farmland in Northeast China. Water 15:869. https://doi.org/10.3390/w15050869

7. Tan Y, Zhao Z (2008) Main methods for determining leaf area index. For Surv Plan 33(3):4. https://doi.org/10.3969/j.issn.1671-3168.2008.03.012
8. Cheng W, Pan K, et al. (2010) Research progress on measurement methods of Leaf Area Index (LAI). Sichuan For Sci Technol 31(3):5. https://doi.org/10.3969/j.issn.1003-5508.2010.03.007
9. Wu W, Hong T, et al. (2007) Research progress on ground measurement methods of leaf area index. J Huazhong Agric Univ 26(2):6. https://doi.org/10.3321/j.issn:1000-2421.2007.02.031
10. Fei L, Mistele B, et al. (2014) Reflectance estimation of canopy nitrogen content in winter wheat using optimised hyperspectral spectral indices and partial least squares regression. Eur J Agron (52-Pt.B). https://doi.org/10.1016/j.eja.2013.09.006
11. Wu L, Tao Q (1999) Study on rice chlorophyll meter diagnosis and nitrogen application method. J Zhejiang Univ (Agric Life Sci Ed) (2):135–138. https://doi.org/10.3321/j.issn:1008-9209.1999.02.006
12. Shen Z, Wang K, et al. (2002) Preliminary report on chlorophyll meter diagnosis of different rice varieties at different nitrogen levels. Sci Technol Bull. https://doi.org/10.3969/j.issn.1001-7119.2002.03.001
13. Wu L, Tao Q (1999) Study on rice chlorophyll meter diagnosis and nitrogen application method. J Zhejiang Agric Univ 25(2):4. https://doi.org/10.3321/j.issn:1008-9209.1999.02.006
14. Tang Y, Wang R, et al. (2003) Study on rapid determination of barley nitrogen nutrition status by hyperspectral and chlorophyll meter. J Triticeae Crop 23(01). https://doi.org/10.7606/j.issn.1009-1041.2003.01.025
15. Zhu X, Sheng H, et al. (2005) Preliminary study on prediction of wheat leaf chlorophyll and nitrogen content using SPAD value. J Triticeae Crop 25(02). https://doi.org/10.7606/j.issn.1009-1041.2005.02.057
16. Li Z, Zhang Y, et al. (2005) Application of chlorophyll meter in nitrogen nutrition diagnosis of summer maize. J Plant Nutr Fertil Sci 11(6):764. https://doi.org/10.11674/zwyf.2005.0609
17. Li Z, Liu H, et al. (2003) Study on application of chlorophyll meter in diagnosis of winter wheat nitrogen nutrition status. J Plant Nutr Fertil Sci 9(4):5. https://doi.org/10.3321/j.issn:1008-505X.2003.04.004
18. Olav HB, Solhaug KA (1998) Effect of irradiance on chlorophyll estimation with the Minolta SPAD-502 leaf chlorophyll meter. Ann Bot (3):389–392. https://doi.org/10.1006/anbo.1998.0683
19. Lukina EV, Freeman KW, et al. (2001) Nitrogen fertilization optimization algorithm based on in-season estimates of yield and plant nitrogen uptake. J Plant Nutr 24(6):885–898. https://doi.org/10.1081/PLN-100103780
20. Mangiafico SS, Guillard K (2007) Cool-Season turfgrass color and growth calibrated to leaf nitrogen.Crop Science 47(3). https://doi.org/10.2135/cropsci2006.04.0259
21. He D (2012) Research on estimation of rice chlorophyll content based on large margin nearest neighbor classification. https://doi.org/10.7666/d.Y2235136
22. Goran K, Aristotelis T, et al. (2019) A new low-cost portable multispectral optical device for precise plant status assessment. Comput Electron Agric 162:300–308. ISSN 0168–1699. https://doi.org/10.1016/j.compag.2019.04.021
23. Zhao C, Liu L, et al. (2004) Development and application of normalized difference vegetation index instrument. Opt Tech 30(3):4. https://doi.org/10.3321/j.issn:1002-1582.2004.03.023
24. Ni J, Yao X, et al. (2013) Design and experiment of portable crop growth monitoring and diagnosis instrument. J Agric Eng (06):158–164. https://doi.org/10.3969/j.issn.1002-6819.2013.06.019
25. Ding Y, Jiang Y, et al. (2020) Measurement method for Height-Independent vegetation indices based on an active light source. Sensors 20(7):1830. https://doi.org/10.3390/s20071830
26. Karthikeyan L, Chawla ILA, et al. (2020) A review of remote sensing applications in agriculture for food security: crop growth and yield, irrigation, and crop losses. J Hydrol 586:124905. ISSN 0022–1694. https://doi.org/10.1016/j.jhydrol.2020.124905
27. Li Y (1993) Dynamic monitoring and Yield estimation of winter wheat using meteorological satellite remote sensing. Meteorol Press

28. Muhammad AH, Yang M (2019) A rapid monitoring of NDVI across the wheat growth cycle for grain yield prediction using a multi-spectral UAV platform. Plant Sci. 282:95–103. ISSN 0168–9452. https://doi.org/10.1016/j.plantsci.2018.10.022
29. Guan S, Fukami K, et al. (2019) Assessing correlation of High-Resolution NDVI with fertilizer application level and Yield of rice and wheat crops using small UAVs. Remote Sens 11(2):112. https://doi.org/10.3390/rs11020112
30. Song X, Wang J, et al. (2004) Study on soil nitrogen supply and variable fertilization effects on winter wheat growth using aerial imaging spectral data. Trans Chin Soc Agric Eng 20(4):5. https://doi.org/10.3321/j.issn:1002-6819.2004.04.010
31. Liu H (1997) Introduction to the application status of agricultural remote sensing technology in the united states. Remote Sens Land Resour 03:56–60. https://doi.org/10.6046/gtzyyg.1997.03.10
32. Zhang R, Li L, et al. (2021) Monitoring the growth status of rapeseed using unmanned aerial vehicle image data. GeomatS Inf Sci Wuhan Univ (S01):046. https://doi.org/10.14188/j.2095-6045.2021170
33. Liu Y, Gong Y, et al. (2020) Monitoring rapeseed growth status using unmanned aerial vehicle remote sensing based on Multi-Temporal NDVI and abundance comprehensive analysis. J Wuhan Univ: Inf Sci Ed (2):8. https://doi.org/10.13203/j.whugis20180161
34. Maimaitijiang M, Sagan V, et al. (2020) Crop monitoring using Satellite/UAV data fusion and machine learning. Remote Sens 12(9):1357. https://doi.org/10.3390/rs12091357
35. Duan B, Fang S, et al. (2019) Remote estimation of rice yield with Unmanned Aerial Vehicle (UAV) Data and Spectral Mixture Analysis. Front Plant Sci 10:204. https://doi.org/10.3389/fpls.2019.00204
36. Wang P (2017) Application of Near-Infrared spectroscopy detection technology in agriculture and food analysis. Agric Prosper Consult 4:1. https://doi.org/10.3969/j.issn.1003-9902.2017.04.055
37. Liu J, Li X, et al. (2007) Research progress on data processing methods based on Near-Infrared spectroscopic analysis. In: Proceedings of the annual meeting of the Chinese society of agricultural engineering, ConferenceArticle/5aa1bba8c095d72220940114
38. Zhu S (2003) Research on quality inspection method of Near-Infrared spectroscopy. Diss, China Agric Univ, https://doi.org/10.7666/d.y558415.
39. Chen B, Zou X, et al. (2008) PCA combined with mahalanobis distance method to eliminate Near-Infrared abnormal samples. J Jiangsu Univ: Nat Sci Ed 29(4):4. CNKI:SUN:JSLG.0.2008–04–003
40. Fang Y, Kong C, Lan, et al. (2006) Application of wavelet transform to noise removal and baseline correction of spectra. Opt Precis Eng 14(6):1088–1092. https://doi.org/10.3321/j.issn:1004-924X.2006.06.027.
41. Zhang L, Zhao L, et al. (2005) Application of MAXR regression method in Near-Infrared spectroscopic quantitative analysis and optimal wavelength selection. Spectrosc Spectr Anal 25(8):3. https://doi.org/10.3321/j.issn:1000-0593.2005.08.016
42. Jian Z, Abdul MM (2023) A novel non-integer order Savitzky–Golay derivative function of visible and near-infrared spectra for improving prediction accuracy of phosphorus in pig manure. Chemom Intell Lab Syst 240:104917. ISSN 0169–7439.https://doi.org/10.1016/j.chemolab.2023.104917.
43. Zhai Y, Zhou L, et al. (2023) Application of Visible/Near-Infrared spectroscopy and hyperspectral imaging with machine learning for High-Throughput plant heavy metal stress phenotyping: a review, plant phenomics. https://doi.org/10.34133/plantphenomics.0124, 5
44. Lu N, Zhou J, et al. (2019) Improved estimation of aboveground biomass in wheat from RGB imagery and point cloud data acquired with a low-cost unmanned aerial vehicle system. Plant Methods 15:17. https://doi.org/10.1186/s13007-019-0402-3
45. Sun J, Yang L, et al. (2021) Using spectral reflectance to estimate the leaf chlorophyll content of maize inoculated with arbuscular mycorrhizal fungi under water stress. Front Plant Sci 12:646173. https://doi.org/10.3389/fpls.2021.646173

46. Xiang Y, An J, et al. (2023) Estimation of soybean growth parameters and yield based on UAV multispectral remote sensing. Trans Chin Soc Agric Mach 54(8):230–239. https://doi.org/10.6041/j.issn.1000-1298.2023.08.022
47. Song X (2023) Research on UAV remote sensing monitoring of main growth parameters of wheat. Diss, Yangzhou Univ, https://doi.org/10.27441/d.cnki.gyzdu.2023.002268
48. Liu C (2023) Research on inversion of cotton leaf SPAD value and LAI based on UAV multispectral images. Diss, Tarim Univ, https://doi.org/10.27708/d.cnki.gtlmd.2023.000183
49. Zhao S, Wang G, Hu, et al. (2024) Estimation of cotton growth Pparameters and yield based on UAV multispectral remote sensing. Trans Chin Soc Agric Mech 45(2):227–234. https://doi.org/10.13733/j.jcam.issn.2095-5553.2024.02.033
50. Xie Q, Jadu D, et al. (2019) Retrieval of crop biophysical parameters from Sentinel-2 remote sensing imagery. Int J Appl Earth Obs Geoinformation 80:187–195. ISSN 1569–8432. https://doi.org/10.1016/j.jag.2019.04.019
51. Yan T, Xu W, et al.(2021) Combining multi-dimensional Convolutional Neural Network (CNN) With visualization method for detection of aphis gossypii glover infection in cotton leaves using hyperspectral imaging. Front Plant Sci 12:604510. https://doi.org/10.3389/fpls.2021.604510
52. Luo S (2019) Comparison of the retrieving precision of potato leaf area index derived from several vegetation indices and spectral parameters of the continuum removal method. Eur J Remote Sens 52(1):155–168. https://doi.org/10.1080/22797254.2019.1572460
53. Zheng H, Cheng T, et al. (2019) Improved estimation of rice aboveground biomass combining textural and spectral analysis of UAV imagery. Precision Agric 20:611–629. https://doi.org/10.1007/s11119-018-9600-7

Chapter 3
Greenhouse Phenotyping Measurement Techniques and Systems: A Review

Tian Gao, Wenyi Sheng, Zhao Zhang, Han Li, and Man Zhang

Abstract Greenhouse phenotypic measurement technology and systems are of great significance in modern plant scientific research and agricultural production. This paper reviews the current development status and future prospects of key measurement technologies and typical systems. First, the three key technologies of three-dimensional measurement, multispectral measurement and thermal infrared imaging are discussed in detail, and their applications and advantages in capturing plant morphological characteristics, health status assessment and environmental adaptability research are analyzed. Secondly, two typical greenhouse phenotyping measurement systems are introduced: PlantEye and FieldScanalyzer. The PlantEye system, developed by Phenospex, is good at high-resolution three-dimensional imaging and multispectral measurement, and is suitable for fine morphological research and physiological characteristics analysis; the FieldScanalyzer system, developed by LemnaTec, has high-throughput measurement and multispectral imaging capabilities, and is suitable for large-scale Rapid acquisition and analysis of large-scale phenotypic data. Finally, the future technological development direction is prospected, emphasizing the key role of software and hardware innovation in improving the performance of greenhouse phenotyping measurement systems.

Keywords Greenhouse phenotyping measurements · 3D measurements · Multispectral measurements · Thermal imaging systems · Plant science

---

T. Gao · W. Sheng (✉) · Z. Zhang · H. Li · M. Zhang
College of Information and Electrical Engineering, China Agricultural University, Beijing 100083, China
e-mail: wenyi.sheng@cau.edu.cn

Key Laboratory of Smart Agriculture System Integration, Ministry of Education, China Agricultural University, Beijing 100083, China

Y. Yun et al. (eds.), *Advanced Sensing and Robotics Technologies in Smart Agriculture*, Smart Agriculture 9, https://doi.org/10.1007/978-981-97-6441-9_3

3.1 Introduction

Greenhouse phenotyping technologies and systems are a series of technologies and systems used to measure and analyze the growth, morphological, physiological and biochemical characteristics of plants in a greenhouse environment [1]. They combine sensor technology, imaging technology, automation systems and data processing methods to enable non-destructive or slightly intrusive measurement and monitoring of plants to achieve efficient data collection and analysis [2]. The primary goal of greenhouse phenotyping technologies and systems is to understand the phenotypic characteristics of plants under various growth conditions, including but not limited to growth rate, leaf area, root morphology, leaf pigment content, and biomass accumulation [3]. By measuring and analyzing these characteristics, scientists can gain a deeper understanding of the response mechanism of plants to environmental factors (such as light, temperature, moisture, etc.), as well as the physiological and biochemical changes of plants during their growth and development [4]. Greenhouse phenotyping measurement technologies and systems usually involve various sensors and equipment, such as high-resolution cameras, lidar scanners, infrared thermal imaging cameras, gas analyzers, soil analyzers, etc. These devices enable comprehensive monitoring and measurement of plants while maintaining stability and control of the greenhouse environment [5].

Greenhouse phenotyping technologies and systems play a crucial role in modern agricultural research. The greenhouse environment provides a controlled growth condition that allows researchers to more accurately measure and analyze plant growth, development, and physiological characteristics [6]. Greenhouse phenotyping measurement technology and systems use various sensors, imaging equipment and automation systems to measure and record plant morphology, structure, growth rate, physiological conditions, etc. non-destructively or slightly disturbingly, providing plants with It provides scientists and agricultural scientists with powerful tools to understand plant growth and adaptation [7].

The introduction and application of greenhouse phenotypic measurement technology and systems have had a profound impact on modern agriculture. They provide an efficient means to evaluate the growth status and biological characteristics of crops, providing agricultural scientists and farmers with more accurate and timely information, thereby helping them better manage crop cultivation and production processes [8]. By conducting detailed monitoring and analysis of plant growth processes, greenhouse phenotypic measurement technology and systems provide important data and tools for crop improvement and adaptive breeding, helping to breed more stress-resistant and high-yielding plants of crop varieties [9]. In addition, greenhouse phenotypic measurement technology and systems also provide digital and automated solutions for agricultural production, improve production efficiency and resource utilization, and help agriculture move towards intelligence and sustainability [10].

3.2 Key Measurement Techniques

3.2.1 3D measurement

3D measurement technology is a method that uses equipment such as laser scanning or stereo cameras to measure and analyze the three-dimensional shape and structure of plants [11]. In greenhouse phenotyping measurements, 3D measurement technology can provide more accurate and comprehensive plant morphological characteristics. Using three-dimensional imaging, the three-dimensional information of the observed object can be obtained and viewed in relevant software. Compared with traditional two-dimensional imaging methods, three-dimensional imaging can more intuitively display physical information such as the volume, height, position, and structure of objects. This technology is widely used in fields such as model forest measurement, cultural relics protection, case investigation, urban planning, medical technology, military and animation art creation [12–17]. Three-dimensional imaging technology is divided into two types: contact and non-contact. Contact three-dimensional imaging technology may affect the imaging results in some cases due to direct contact with the object being measured. For example, when the surface of the object being measured is soft, direct contact will change its appearance, thus affecting imaging accuracy [18]. Non-contact 3D imaging technology is widely adopted to meet the demands that contact-based techniques cannot fulfill [19, 20]. This technology acquires three-dimensional information of objects indirectly, without direct contact. Non-contact 3D imaging technology includes active and passive types. Active 3D imaging technology typically uses lasers or other energy sources to emit light or energy towards the target object, and then generates a 3D model by measuring the reflection or scattering of the light or energy on the object's surface. Passive 3D imaging technology, on the other hand, generates a 3D model by measuring the light or energy emitted by the target object. Active 3D imaging technologies include methods such as laser triangulation [21], time-of-flight [22], structured light [23], and Moiré fringe [24]. Structured light technology works by projecting light with specific feature points onto the object and capturing the projection on the object's surface to calculate its three-dimensional information. Passive 3D imaging technologies include methods such as stereo vision, shape from texture [25], and shape from shading [26]. Detailed classifications of 3D imaging technologies can be referred to in Fig. 3.1.

This article mainly introduces three commonly used methods: laser triangulation, time-of-flight, and stereo vision.

First, laser triangulation typically utilizes a laser line projected onto the object being measured and captures the projection image on the object's surface using a camera. By applying the principles of triangulation, the depth information of the object's surface can be calculated. Finally, by moving the laser line and the object relative to each other, the 3D information of the object's surface can be obtained [27]. Specifically, the laser emits a beam that, after being reflected, enters the camera lens and is captured by the sensor. If there is a height variation on the object's surface,

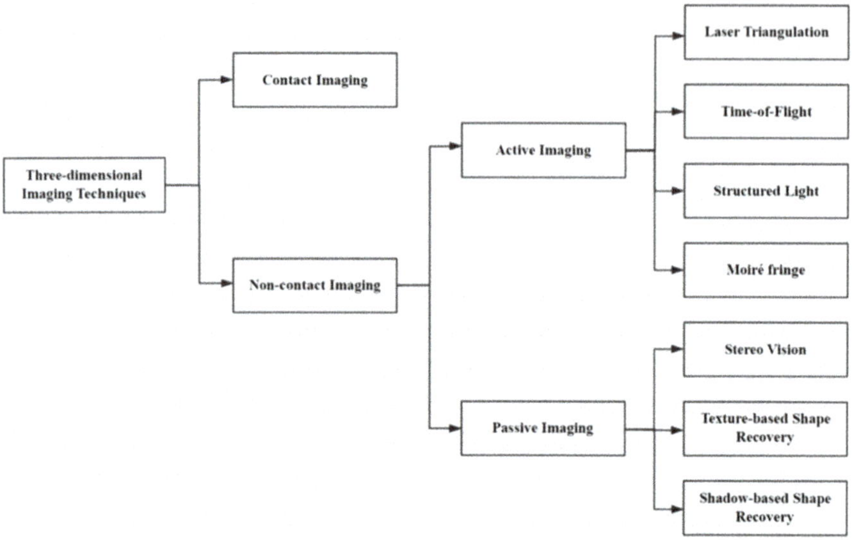

Fig. 3.1 Hierarchical classification of 3D imaging techniques: contact vs. non-contact methods

such as an increase from the blue plane to the red plane with a height y, the reflected laser enters the sensor surface, creating a corresponding distance difference d. As the height y increases, the distance difference d also increases. Common types of lasers include point lasers and line lasers.

The time-of-flight (TOF) method calculates the distance from an object to a sensor by measuring the time it takes for light projected by the sensor to be reflected back. Common TOF depth cameras on the market include Microsoft's Kinect v2 sensor, MESA Imaging's Swiss Ranger 400, and Sony's DepthSense sensors. As shown in Figs. 3.1, 3.2, 3.3 and 3.4. Microsoft's Kinect v2 sensor contains three cameras: an RGB camera, an infrared (IR) camera, and an IR projector, from left to right.

Humans have two eyes positioned on either side of their head, allowing them to observe the same object from different angles, resulting in slightly different images seen by each eye. This difference is called parallax. The greater the parallax, the closer the object is to the eyes; the smaller the parallax, the farther the object is from the eyes.

Fig. 3.2 Schematic representation of the laser triangulation imaging principle

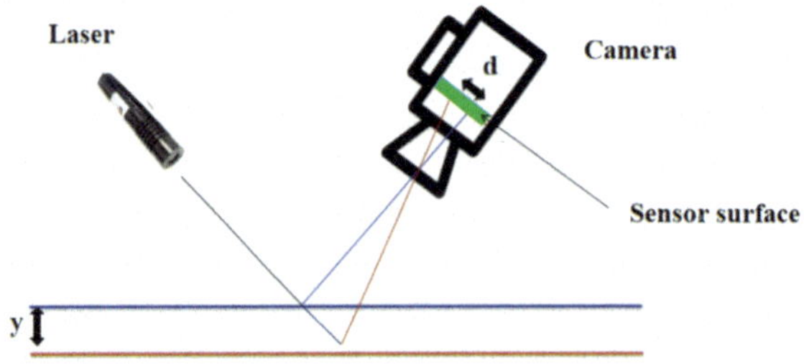

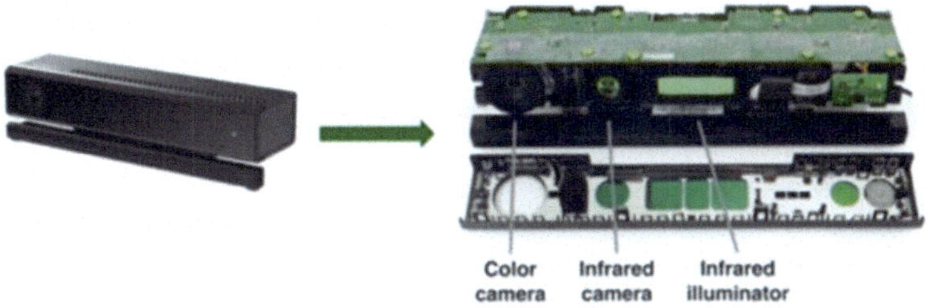

Fig. 3.3 Internal structure and components of the Microsoft Kinect v2 sensor

Fig. 3.4 Zed 2k stereo camera: a crucial component in real-time 3d reconstruction systems

Stereo vision utilizes the parallax difference between human eyes to produce a sense of depth, enabling us to perceive three-dimensional space. A commonly used stereo vision method is binocular stereo vision. The basic approach involves capturing two slightly different images from the left and right angles using two cameras, then pairing these images to generate a sense of depth. A binocular camera consists of two cameras with their lens planes set at a certain angle to each other. A larger angle improves the final 3D imaging effect, but if the angle is too large, the overlapping area of the images decreases, affecting the accuracy of image matching. Therefore, choosing an appropriate angle is crucial when designing binocular cameras. A commonly used product for binocular stereo vision technology is the ZED 2 K Stereo Camera by STEROLABS.

Using 3D measurement technology, the precise three-dimensional morphology and structure of plants, such as leaf shapes and root systems, can be captured. This data can be used to analyze plant growth characteristics, distribution patterns, and associations with environmental factors, helping to understand the plant growth and development process [28]. Additionally, it can accurately measure plant height, root depth, and distribution, which is significant for studying plant growth rates, physiological states, and interactions with the soil environment [1]. 3D measurement technology can also analyze the spatial structure of plant communities, including parameters such as plant height distribution, coverage, and density [29].

3.2.2 *Multispectral Measurements*

Multispectral measurement is a technique that uses multiple wavelengths of light to measure and analyze plant characteristics [30]. In greenhouse phenotyping, multispectral measurement can provide information on the reflection, absorption, and transmission of plants under different spectral bands, revealing their growth status, physiological activity, and interactions with environmental factors. This technology can be implemented through various devices, including spectroradiometers, spectral cameras, and spectrometers [31]. These devices can simultaneously measure spectra across multiple wavelength ranges and convert them into numerical data for analysis. Through multispectral measurement, the reflectance spectra of plants under different bands can be obtained, allowing the study of biological characteristics such as photosynthesis, chlorophyll content, and leaf health [32].

There are three methods for acquiring multispectral imaging data [16]. As shown in the figure below, light from a specific spectral band emitted by the light source towards the object being measured is received by a detector. Based on how the light is received by the detector, multispectral imaging can be categorized into point scanning, line scanning, and area scanning methods. In point scanning, the spectral information of one point is recorded at a time. In line scanning, the spectral information of a line, which consists of a row or column of points, is recorded at a time. In area scanning, the spectral information of an entire area is recorded sequentially, eventually obtaining the spectral information for the entire area (Fig. 3.5).

Multispectral measurement can be used to monitor the growth status and physiological changes of plants, including growth rate, leaf area, and dry weight. It can also assess the health status of plants, including chlorophyll content, leaf moisture content, and leaf nitrogen content. Additionally, multispectral measurement can monitor plant diseases and pest infestations, such as disease spots, pest damage, and leaf damage. By analyzing the spectral reflection characteristics of plants under different bands, it is possible to detect diseases and pest infestations in a timely manner, providing scientific evidence for pest and disease control [33, 34].

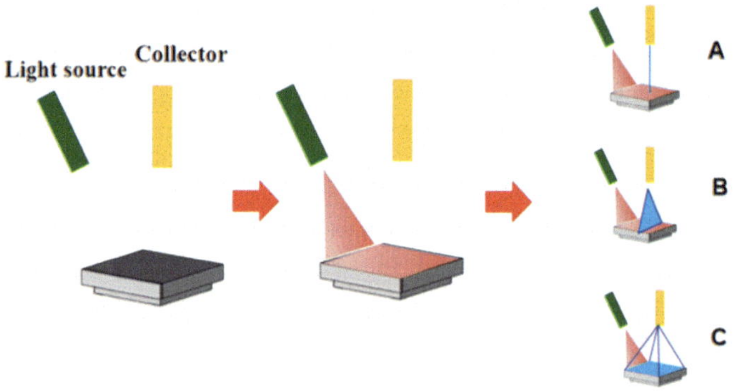

Fig. 3.5 Workflow of three multispectral imaging methods for material characterization

3.2.3 *Infrared Thermography Measurement*

Thermal infrared imaging is a technology that measures the temperature distribution on the surface of objects using infrared radiation [35]. In greenhouse phenotyping, thermal infrared imaging can provide information on the temperature distribution of plants under different times and environmental conditions, revealing their growth status, physiological activity, and interactions with environmental factors [36]. This technology collects infrared radiation data from the surface of objects using devices such as infrared thermal cameras and converts it into temperature distribution images. By analyzing these images, the temperature variations in different parts of the plant can be understood, thereby assessing the plant's growth status and physiological responses [37].

The principle of thermal infrared imaging is that objects emit infrared radiation when their temperature is above absolute zero. In a thermal infrared imaging system, an infrared thermal camera detects this infrared radiation, typically within the wavelength range of 8–14 μm. Based on the intensity of the infrared radiation emitted by the object, its surface temperature can be inferred. Since the intensity of infrared radiation is directly proportional to temperature, surface temperature of objects can be indirectly measured by measuring the intensity of infrared radiation. By processing and analyzing the collected infrared radiation data, a thermal infrared imaging system can generate temperature distribution images of object surfaces. These images are usually presented in the form of thermal maps, where different colors represent different temperature zones. The generated temperature distribution images can be used to analyze the temperature distribution of objects, revealing the temperature variations of plants under different times and environmental conditions, thereby assessing the growth status and physiological responses of plants [37]. The principle is illustrated in the Fig. 3.6.

Thermal infrared imaging can be used to study plant physiological activities such as photosynthesis, transpiration, and stomatal movement. By monitoring temperature changes in plant leaves, insights into plant physiological responses and regulatory mechanisms can be gained. It can also be utilized to monitor plant growth status and growth rate. Analyzing temperature distribution in different parts of plants allows

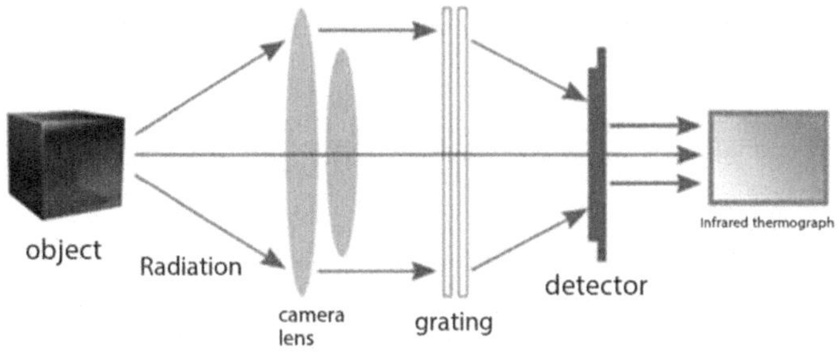

Fig. 3.6 Schematic diagram of the thermal infrared imaging process

for the assessment of plant growth conditions and trends. Additionally, it can be used to evaluate the moisture and nutritional status of plants. Since plant moisture and nutritional status affect their surface temperature distribution, thermal infrared imaging can indirectly assess the moisture and nutritional status of plants [38, 39].

In this section, three commonly used greenhouse phenotyping techniques were introduced: 3D measurement, multispectral measurement, and thermal infrared imaging. These techniques each have their own characteristics and can provide rich information about plant features, aiding in a deeper understanding of plant growth, development, and adaptability. Firstly, 3D measurement technology provides comprehensive plant morphological characteristics by capturing the three-dimensional morphology and structure of plants, such as leaf area, plant height, and root system structure. Its high precision and comprehensiveness make it an important tool for studying plant growth and genetic breeding. Secondly, multispectral measurement technology utilizes spectra from multiple wavelengths for plant feature measurement and analysis, enabling assessment of plant physiological status, leaf health, and photosynthetic efficiency, among others. Its non-contact and high-throughput characteristics make it widely applicable in efficient plant phenotyping. Lastly, thermal infrared imaging technology reveals plant temperature variations under different environmental conditions by monitoring the temperature distribution on the plant surface, aiding in the assessment of plant growth status, physiological activities, and adaptability.

3.3 Typical Greenhouse Phenotype Measurement System

In the field of plant science and agricultural production, the application of greenhouse phenotyping systems is increasingly receiving attention. These systems not only provide efficient and accurate plant feature data but also offer important technical support for plant research, genetic breeding, and agricultural production. This article will focus on introducing two typical greenhouse phenotyping systems, PlantEye and FieldScanalyzer.

3.3.1 PlantEye

PlantEye is a greenhouse phenotyping system based on three-dimensional imaging technology, as shown in the figure below. Developed by Phenospex company, this system aims to achieve precise and efficient measurement and analysis of plant morphology, structure, and physiological characteristics. PlantEye utilizes high-resolution 3D imaging sensors and advanced image processing algorithms to accurately capture and reconstruct the microstructure and details of plants [40] (Fig. 3.7).

Fig. 3.7 PlantEye device deployed in a greenhouse for high-precision crop analysis

The PlantEye system employs omnidirectional imaging technology, enabling comprehensive and three-dimensional measurement and observation of plants from different angles and directions. This omnidirectional measurement capability allows the system to capture the morphological and structural characteristics of plants at different spatial scales, providing a more comprehensive data foundation for plant phenotypic analysis. Additionally, the system is equipped with an automatic control system, enabling automated acquisition, imaging, and data processing of plants. Users can achieve automated operation of the measurement process through preset parameters and workflows, thereby improving measurement efficiency and accuracy. Moreover, the PlantEye system is equipped with advanced data processing and analysis software, capable of visualizing and analyzing the collected data [41, 42].

The PlantEye system comprises several components, including a sensor module consisting of 3D imaging sensors such as cameras, laser scanners, or other imaging devices. It also includes a mechanical frame used to stabilize and accurately position the sensors and plant samples during the imaging process. Additionally, it is equipped with an automated control system for automating the imaging process, as well as related software for data processing.

The main workflow of the system begins with image acquisition, where the sensors capture comprehensive images of the plant from various angles and directions, obtaining three-dimensional morphological and structural information. Next is data processing, which involves 3D reconstruction of the images, feature extraction, and data analysis, enabling precise measurement and quantitative analysis of the plant's morphology, structure, and physiological characteristics. This is followed by feature extraction, which can identify various morphological, structural, and physiological features of the plant, including leaf area, leaf angle, leaf exposure area, root length, and root volume. Finally, these data can be analyzed for further insights [43]. For example, the figure below shows a 3D plant model created using the PlantEye system (Fig. 3.8).

PlantEye system, as a greenhouse phenotyping system based on three-dimensional imaging technology, finds extensive applications in plant science research, crop genetic breeding, and agricultural production. For instance, the PlantEye system enables real-time monitoring of plant growth processes, tracking growth curves,

Fig. 3.8 Corn plant 3D model generated for phenotypic analysis

Fig. 3.9 Enhanced NDVI 3D model for monitoring plant growth and health

leaf expansion, and root growth characteristics [44]. Utilizing spectral information obtained from 3D models, various spectral indices such as NDVI and PSRI can be computed. The NDVI model (as shown in the figure below) allows the calculation of various parameters associated with plant senescence, such as healthy and unhealthy leaves (Fig. 3.9).

3.3.2 FieldScanalyzer

FieldScanalyzer is an advanced high-throughput phenotyping system developed by LemnaTec, designed for efficient and accurate phenotypic measurement and data analysis of plants in greenhouse or field environments. By integrating multispectral imaging, 3D imaging, and automated control technologies, this system provides comprehensive plant phenotypic feature measurement and analysis capabilities [1] (Fig. 3.10).

The FieldScanalyzer system is equipped with multispectral cameras that can capture images of plants across different spectral ranges, including visible light, near-infrared, and thermal infrared. It can efficiently measure a large number of

Fig. 3.10 Field Scanalyzer system for automated field phenotyping

plant samples simultaneously, making it suitable for large-scale plant phenotyping research and crop genetic breeding projects. The system also features an automated control module that enables automatic positioning, imaging, and data processing of plant samples. Additionally, FieldScanalyzer has 3D imaging capabilities, allowing for the three-dimensional reconstruction and analysis of plant morphology and structure. The system is also equipped with powerful data analysis software that can process, analyze, and visualize the collected images and data [1].

The FieldScanalyzer system is a 3-axis sensor-to-plant phenotyping analysis system. It is designed and constructed based on an industrial gantry crane system. The x-axis is guided along a track system supported by concrete piles driven into the ground to allow for natural drainage. The y-axis is orthogonal to the track and carries a lifting unit with a container of sensing equipment. This setup allows for precise movement of all three axes throughout the year. The sensor box can reach each point in the measurement area according to a predefined schedule with sub-centimeter repeatable high precision. Any camera system will be installed in a separate weatherproof sealed enclosure and mounted on a flexible and adjustable platform within the imaging box. Sensor options include visible light camera modules, near-infrared camera modules, chlorophyll fluorescence dynamics modules, hyperspectral imaging modules, multispectral imaging modules, infrared camera modules, 3D laser scanning modules, and weather and environmental condition sensors [45]. As shown in the Fig. 3.11.

The FieldScanalyzer system has a wide range of applications in plant science research and agricultural production. It can rapidly measure and compare various phenotypic traits such as growth rates, leaf health, photosynthetic efficiency, and disease resistance across different plant varieties. This data can be used to accelerate the breeding process and select varieties with strong adaptability, high yield, and good stress resistance [46]. The multispectral imaging technology of the FieldScanalyzer system can identify early symptoms of plant diseases, such as leaf spots and changes

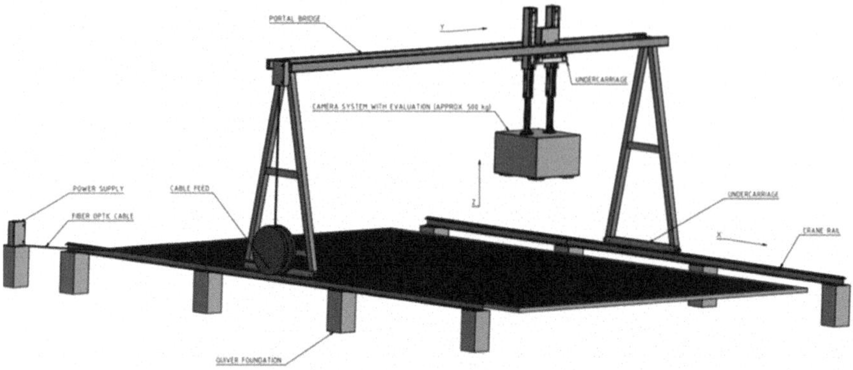

Fig. 3.11 Structural overview of the Field Scanalyzer system for crop phenotyping

in chlorophyll content. By detecting diseases early, agricultural producers can quickly implement appropriate control measures [37].

The FieldScanalyzer system can measure physiological parameters such as leaf area, leaf angle, and canopy structure, aiding the study of physiological processes like photosynthesis, transpiration, and nutrient absorption [7]. Additionally, the FieldScanalyzer system can efficiently measure a large number of plant samples simultaneously, making it suitable for large-scale plant phenotyping studies and genotype–phenotype association research [47].

Both PlantEye and FieldScanalyzer have high-throughput measurement capabilities, allowing for the measurement and analysis of a large number of plant samples in a short time. They both feature multispectral imaging, automated control, and data analysis and visualization.

The difference lies in their focus: PlantEye specializes in high-resolution 3D imaging, capable of precisely capturing the three-dimensional morphology and structural characteristics of plants, making it suitable for detailed morphological studies. While FieldScanalyzer also has 3D imaging capabilities, its strengths lie in its multispectral imaging and high-throughput measurement capabilities, making it ideal for rapid acquisition and analysis of large-scale phenotypic data. PlantEye is typically used for plant phenotyping in greenhouse environments, making it more suitable for monitoring plant growth and analyzing physiological characteristics under controlled conditions. In contrast, FieldScanalyzer is suitable not only for greenhouse environments but also for field phenotyping, offering a broader range of application scenarios.

3.4 Future Outlook

3.4.1 Hardware Innovation

In greenhouse phenotyping systems, hardware innovation is key to improving system performance, accuracy, and applicability. Future hardware innovations will drive the development of greenhouse phenotyping technology in several ways. In the crucial area of sensors, advancements in high-resolution imaging sensors will significantly enhance the accuracy and detail-capturing capability of plant phenotyping measurements. Future sensors will have higher spatial and spectral resolution, enabling the capture of finer structural and physiological changes in plants [30]. For example, new hyperspectral cameras can image across more spectral bands, providing richer information on plant health and nutrition.

To achieve higher precision and speed in 3D imaging technology and quickly generate high-resolution 3D models of plants, the development of light field cameras and LiDAR technology will be crucial [48]. Additionally, integrating multiple sensors into a single system for higher integration is a key direction in hardware development. Future hardware integration will be more advanced, incorporating multispectral cameras, thermal infrared sensors, LiDAR, and environmental sensors into one system, enabling simultaneous acquisition of plant morphology, spectral, and environmental data [46].

The development of automation and robotics technology will also enhance the efficiency and accuracy of greenhouse phenotyping. Future systems will be equipped with automated mobile platforms and robotic arms capable of autonomous navigation, positioning, and measuring plant samples within the greenhouse. These automated devices can operate around the clock, significantly reducing labor requirements and improving the consistency and reliability of data collection [2].

The development of portable and wearable sensors is another direction in hardware advancement. These sensors can be directly attached to plants or carried by researchers for real-time data collection and monitoring. For example, miniature spectral sensors and wireless environmental sensors can enable real-time monitoring and analysis of the plant growth environment, supporting precision agriculture management [49].

3.4.2 Software Innovation

As plant phenotyping technologies continue to develop, software plays an increasingly crucial role in data collection, processing, and analysis. Future software innovations will drive the advancement of greenhouse phenotyping systems in multiple ways. In the critical area of data analysis, future software will rely more on advanced data analysis techniques and machine learning algorithms to handle and analyze vast amounts of plant phenotypic data. Machine learning models can be used to identify and predict plant growth patterns, disease symptoms, and physiological responses. For example, deep learning algorithms can automatically analyze multispectral and 3D imaging data to extract information about plant health and growth characteristics. Through continuous training and optimization, these models will become more accurate and efficient [50].

At the same time, future software will focus more on data integration and interoperability, allowing seamless integration of data from different sources and formats. This includes integrating multispectral imaging, 3D imaging, environmental sensor data, and genomic data to build a comprehensive plant phenotyping database [51]. Future software will offer more powerful real-time data processing capabilities, utilizing cloud computing and distributed computing technologies to achieve real-time processing and analysis of large-scale data (Singh [52]. Additionally, future software will emphasize user-friendliness and customizable design to facilitate user operation. Open platforms and community collaboration will be key directions for future software innovation [53].

3.5 Summarize

Greenhouse phenotyping technology and systems play a crucial role in plant science research and agricultural production. This paper reviews the current major measurement technologies, the development of typical systems, and future prospects. First, regarding key measurement technologies, 3D measurement, multispectral measurement, and thermal infrared imaging have been instrumental in plant phenotyping research. 3D measurement technologies, such as LiDAR and structured light, can accurately capture plant morphological characteristics and growth status, making them suitable for morphological studies and growth monitoring. Multispectral measurement technology, by capturing images in different spectral bands, provides detailed information on plant health and physiological traits, aiding in early disease detection and nutrient analysis. Thermal infrared imaging, by detecting temperature changes on the plant surface, helps researchers understand plant transpiration and water status, and is widely used in studies of environmental adaptability and drought warning.

Secondly, the typical greenhouse phenotyping systems, PlantEye and FieldScanalyzer, each have their own unique features. The PlantEye system, developed by Phenospex, focuses on high-resolution 3D imaging and multispectral measurement, making it suitable for detailed morphological studies. The FieldScanalyzer system, developed by LemnaTec, offers high-throughput measurement and multispectral imaging capabilities, making it suitable for rapid analysis of large-scale phenotypic data. Both systems have different advantages in high-throughput measurement, 3D imaging, and multispectral imaging, allowing researchers to choose the appropriate system based on specific needs.

In the future, greenhouse phenotyping technology will continue to innovate in both software and hardware. On the software side, machine learning and data integration technologies will improve data processing and analysis efficiency. Real-time data processing and user-friendly design will enhance the ease of use of the systems. On the hardware side, the development of high-resolution imaging sensors, advanced 3D imaging technologies, multifunctional sensor integration, and automation and robotics technologies will enhance the accuracy and efficiency of the systems.

References

1. Fiorani F, Schurr U (2013) Future scenarios for plant phenotyping. Annu Rev Plant Biol 64:2–9
2. Walter A, Liebisch F, Hund A (2015) Plant phenotyping: from bean weighing to image analysis. Plant Methods 11:1–11
3. Minervini M, Giuffrida MV, Perata P et al (2017) Phenotiki: An open software and hardware platform for affordable and easy image-based phenotyping of rosette-shaped plants. Plant J 90(1):4–16
4. Montesano FF, Van Iersel MW, Boari F, Cantore V, D'Amato G, Parente A (2018) Sensor-based irrigation management of soilless basil using a new smart irrigation system: effects of set-point on plant physiological responses and crop performance. Agric Water Manag 203:20–29
5. White JW, Conley MM (2013) A flexible, low-cost cart for proximal sensing. Crop Sci 53(4):2–6
6. Tardieu F, Tuberosa R (2010) Dissection and modelling of abiotic stress tolerance in plants. Curr Opin Plant Biol 13(2):6–12
7. Tardieu F, Cabrera-Bosquet L, Pridmore T, Bennett M (2017) Plant phenomics, from sensors to knowledge. Curr Biol 27(15):R770–R783
8. Tuberosa R (2012) Phenotyping for drought tolerance of crops in the genomics era. Front Physiol 3:2–9
9. Reynolds M, Manes Y, Izanloo A, Langridge P (2009) Phenotyping approaches for physiological breeding and gene discovery in wheat. Ann Appl Biol 155(3):9–12
10. Cabrera-Bosquet L, Crossa J, von Zitzewitz J, Serret MD, Luis Araus J (2012) High-throughput phenotyping and genomic selection: the frontiers of crop breeding converge F. J Integr Plant Biol 54(5):2–8
11. Liu Y, Pears N, Rosin PL, Huber P (eds) (2020) 3D imaging, analysis and applications. Springer International Publishing, Cham
12. Yuan GP (2018) Application of 3D laser scanning technology in cultural relics protection. Mine Survey 46(05):93–97
13. Bu LB (2021) 3D reconstruction method of crime scenes based on structured light [Doctoral dissertation, China People's Public Security University]
14. Tang Z, Zhang XC, Cao KB (2010) Research on the application of 3D technology based on Skyline in urban planning. Bull Surv Mapp 398(05):10–12+41
15. Men HY (2008) Research and application of 3D reconstruction technology in medical imaging [Doctoral dissertation, University of Electronic Science and Technology of China]
16. Wu T, Fu ZL (2015) 3D reconstruction technology and its military applications. Defense Technol 36(01):31–34
17. Deng Q (2019) Research on dimensions of 3D animation art creation [Doctoral dissertation, Xi'an Academy of Fine Arts]
18. Ding SW, Zhang XH, Yu QF et al (2017) Review of non-contact 3D reconstruction measurement methods. Laser Optoelectron Progr 54(07):27–41
19. Li JL, Xin QQ, Tian L et al (2014) Review of online size measurement methods for large forgings. New Industrializat 4(01):59–64
20. Xiong LL, Qian D, Li HZ (2013) A review of real-time 3D reconstruction. J Commun Univ China (Sci Technol) 20(06):38–43

21. Wan J, Huang YQ (2006) Research on laser triangulation measurement. J Sanming Univ 04:361–364
22. Lu CQ, Song YZ, Wu YP, Yang MF (2018) Acquisition and error analysis of 3D information based on TOF computational imaging. Infrared Laser Engin 47(10):160–166
23. Su XY, Zhang QC, Chen WJ (2014) Structured light 3D imaging technology. Chin J Lasers 41(02):9–18
24. Li EZ (2017) Key technologies of 3D contour measurement based on fringe projection method [Doctoral dissertation, Xi'an Technological University]
25. Hu DD (2019) Research on 3D reconstruction based on image texture information [Doctoral dissertation, Huazhong University of Science and Technology]
26. Yang XB, Huang XB (2003) Research on reconstruction of wrinkled fabric surface shape based on shape from shading. J Image Graph 02:63–69
27. Wang ZS, Jia YP, Zhang J, Wang RH (2021) Application of multispectral imaging technology in botanical research. Bull Botany 56(04):1–8
28. Paulus S, Schumann H, Kuhlmann H, Léon J (2014) High-precision laser scanning system for capturing 3D plant architecture and analysing growth of cereal plants. Biosys Eng 121:1–11
29. Hartley ZKJ, Jackson AS, Pound M et al (2021) GANana: unsupervised domain adaptation for volumetric regression of fruit. Plant Phenom 2021:1–11
30. Sankaran S, Khot LR, Espinoza CZ et al (2015) Low-altitude, high-resolution aerial imaging systems for row and field crop phenotyping: a review. Eur J Agron 70:112–123
31. Liebisch F, Kirchgessner N, Schneider D et al (2015) Remote, aerial phenotyping of maize traits with a mobile multi-sensor approach. Plant Methods 11:9
32. Shakoor N, Lee S, Mockler TC (2017) High throughput phenotyping to accelerate crop breeding and monitoring of diseases in the field. Curr Opin Plant Biol 38:184–192
33. Gonzalez-Dugo V, Zarco-Tejada P, Nicolás E et al (2013) Using high resolution UAV thermal imagery to assess the variability in the water status of five fruit tree species within a commercial orchard. Precision Agric 14:660–678
34. Ouf NS (2018) A review on the relevant applications of machine learning in agriculture. Methods 6(8):54–65
35. Jones HG, Stoll M, Santos T et al (2002) Use of infrared thermography for monitoring stomatal closure in the field: application to grapevine. J Exp Bot 53(378):2249–2260
36. Wahabzada M, Mahlein AK, Bauckhage C et al (2015) Metro maps of plant disease dynamics—automated mining of differences using hyperspectral images. PLoS ONE 10(1):e0116902
37. Mahlein AK (2016) Plant disease detection by imaging sensors–parallels and specific demands for precision agriculture and plant phenotyping. Plant Dis 100(2):241–251
38. Mutka AM, Bart RS (2015) Image-based phenotyping of plant disease symptoms. Front Plant Sci 5:734
39. Elsayed S, Elhoweity M, Ibrahim HH et al (2017) Thermal imaging and passive reflectance sensing to estimate the water status and grain yield of wheat under different irrigation regimes. Agric Water Manag 189:98–104
40. Li D, Quan C, Song Z et al (2021) High-throughput plant phenotyping platform (HT3P) as a novel tool for estimating agronomic traits from the lab to the field. Front Bioengin Biotechnol 8:623123
41. Behmann J, Steinrücken J, Plümer L (2014) Detection of early plant stress responses in hyperspectral images. ISPRS J Photogramm Remote Sens 93:98–111
42. Busemeyer L, Mentrup D, Möller K et al (2013) BreedVision—A multi-sensor platform for non-destructive field-based phenotyping in plant breeding. Sensors 13(3):2830–2847
43. Chéné Y, Rousseau D, Lucidarme P et al (2012) On the use of depth camera for 3D phenotyping of entire plants. Comput Electron Agric 82:122–127
44. Klukas C, Chen D, Pape JM (2014) Integrated analysis platform: an open-source information system for high-throughput plant phenotyping. Plant Physiol 165(2):506–518
45. Chawade A, van Ham J, Blomquist H et al (2019) High-throughput field-phenotyping tools for plant breeding and precision agriculture. Agronomy 9(5):258

46. Araus JL, Cairns JE (2014) Field high-throughput phenotyping: The new crop breeding frontier. Trends Plant Sci 19(1):52–61
47. Yang W, Duan L, Chen G et al (2013) Plant phenomics and high-throughput phenotyping: accelerating rice functional genomics using multidisciplinary technologies. Curr Opin Plant Biol 16(2):180–187
48. Dhondt S, Wuyts N, Inzé D (2013) Cell to whole-plant phenotyping: the best is yet to come. Trends Plant Sci 18(8):428–439
49. Sankaran S, Mishra A, Ehsani R et al (2010) A review of advanced techniques for detecting plant diseases. Comput Electron Agric 72(1):1–13
50. Pound MP, Atkinson JA, Townsend AJ et al (2017) Deep machine learning provides state-of-the-art performance in image-based plant phenotyping. Gigascience 6(10):1–10
51. Ubbens JR, Stavness I (2017) Deep plant phenomics: a deep learning platform for complex plant phenotyping tasks. Front Plant Sci 8:1190
52. Singh A, Ganapathysubramanian B, Singh AK et al (2016) Machine learning for high-throughput stress phenotyping in plants. Trends Plant Sci 21(2):110–124
53. Fahlgren N, Gehan MA, Baxter I (2015) Lights, camera, action: high-throughput plant phenotyping is ready for a close-up. Curr Opin Plant Biol 24:93–99

Chapter 4
On-Site Intelligent Grading System for Fruits and Vegetables: Case Studies on Tomato and Apple

Yuliang Yun, Zhiyong Ma, Xueke An, and Dehao Li

Abstract With the improvement of people's living standards, fruits and vegetables have gradually become an indispensable part of human daily diet. Eating more fruits and vegetables can maintain good health and prevent diseases. Fruits and vegetables with regular shapes and bright colors are more popular among people, so the needs of consumers can be met through fruit and vegetable grading. In addition, fruit and vegetable grading can also help improve product quality and market competitiveness, reduce costs, and ensure the quality of fruits and vegetables. Traditional manual fruit and vegetable grading suffers from high labor costs, low grading accuracy, and susceptibility to subjective factors. On-site intelligent grading system for fruits and vegetables at the production site is comparatively faster and more efficient than manual grading. It boasts higher precision, significantly reduces labor costs, and greatly enhances production efficiency and economic benefits. This review primarily discusses the research progress of intelligent grading equipment for fruits and vegetables at the production site. Specifically, this paper provides an overview of On-site Intelligent Grading System for Fruits and Vegetables at the production site, including the basic concepts and structural components of grading equipment. Additionally, this paper focuses on the application of machine vision technology in fruit and vegetable grading. It briefly outlines machine vision technology and reviews its research outcomes in fruit and vegetable grading, primarily focusing on common examples such as tomatoes and apples. Finally, the paper provides an outlook on future research prospects. This review primarily emphasizes the existing achievements and immense potential of machine vision technology in intelligent grading of fruits and vegetables.

Keywords Fruit and vegetable grading · Intelligent system · Machine vision · External detection · Image processing

Y. Yun (✉) · Z. Ma · X. An · D. Li
College of Mechanical and Electrical Engineering, Qingdao Agricultural University, Qingdao 266109, China
e-mail: ylyun1981@163.com

Y. Yun et al. (eds.), *Advanced Sensing and Robotics Technologies in Smart Agriculture*, Smart Agriculture 9, https://doi.org/10.1007/978-981-97-6441-9_4

4.1 Introduction

Fruits and vegetables serve as crucial sources of dietary nutrients, encompassing rich reservoirs of carbohydrates, vitamins, and minerals, playing pivotal roles in fostering both healthy dietary patterns and balanced lifestyles [1]. According to data from the Food and Agriculture Organization (FAO) of the United Nations, global fruit production increased by 63% between 2000 and 2021, reaching 9.33 billion tons in 2021. Among these figures, apple production stood at 96 million tons, trailing behind bananas (135 million tons) and watermelons (100 million tons). Simultaneously, vegetable production also witnessed a 71% increase, reaching 1.17 billion tons in 2022. In that year, tomatoes topped the production charts, totaling 186 million tons [2]. Apples and tomatoes are among the most extensively cultivated fruits and vegetables globally. Their varied shapes, sizes, and colors pose significant challenges in the grading process due to their diversity. Therefore, this paper focuses on tomatoes and apples as the primary research subjects, providing a comprehensive review of their research outcomes. This aims to offer valuable insights and experiences for other similar agricultural products. The global production figures for tomatoes and apples from 2000 to 2022 are illustrated in Fig. 4.1.

As living standards improve, there is an increasing demand for higher quality and better appearance of fruit and vegetable products. Consumers are placing greater emphasis on the appearance, taste, and overall quality of the products [3]. Therefore, it is imperative for fruit and vegetable producers to implement grading processes at their origin to enhance the appearance, quality, and consistency of their products, thus meeting market demands [4].

Fruit and vegetable origin grading inspection can be categorized into internal and external inspections. Internal inspection involves assessing and evaluating the

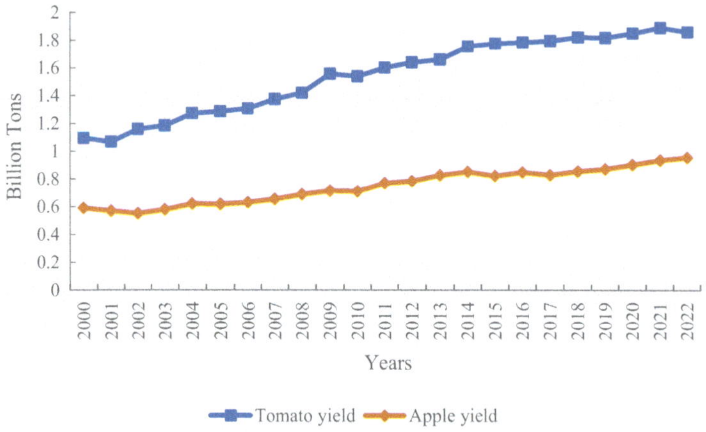

Fig. 4.1 Global tomato and apple production. *Source* FAO. 2023. Production: Crops and livestock products. In: FAOSTAT. Rome. [Cited December 2023]. https://www.fao.org/faostat/en/#data/QCL

internal quality and characteristics of fruits and vegetables, such as sweetness, hardness, ripeness, and other indicators [5–7]. External inspection involves the evaluation and assessment of the outward appearance, size, shape, color, defects, and other characteristics of fruits and vegetables [8]. Consumers often judge the freshness, ripeness, and quality of fruits and vegetables based on their appearance. Fruits and vegetables with a pleasing appearance are more likely to be favored by consumers. Therefore, fruit and vegetable producers typically focus on grading based on external characteristics. However, for internal quality grading of fruits and vegetables, it may be necessary to utilize specialized equipment and techniques for detection and evaluation.

Traditionally, the grading of fruits and vegetables has heavily relied on manual labor.

In the early stages, farmers relied on visual inspection to assess the quality and size of fruits and vegetables, followed by manual sorting and grading. However, this method often faced challenges such as subjective biases from inspectors, high labor costs, increased error rates due to fatigue, lack of real-time data recording and analysis, and difficulties in managing large-scale production [9]. As depicted in Fig. 4.1, the global production of tomatoes and apples nearly doubled over the span of 20 years. This increase has consequently led to significant challenges and increased workloads in manual sorting and grading processes. Against this backdrop, research into On-site intelligent grading system for fruits and vegetables has propelled the transition from traditional manual sorting to automated grading and sorting processes. This advancement offers new solutions to the various challenges encountered in fruit and vegetable grading. Initially, the emergence of mechanized grading equipment introduced the use of transmission systems and mechanical structures for sorting fruits and vegetables. These mechanical devices enabled simple classification of fruits and vegetables based on physical characteristics such as size and weight, thereby enhancing grading efficiency. However, they still encountered issues such as low accuracy in classification and limited applicability. With the rapid development of computer technology and electronic engineering, automated grading equipment has encountered new opportunities for advancement. The application of electronic sensors enables grading equipment to acquire additional information such as color, texture, and other attributes, thereby improving the accuracy of classification. The emergence of computer vision technology has brought about revolutionary changes in fruit and vegetable grading. By utilizing technologies such as cameras and image processing algorithms, it is possible to achieve high-precision detection and analysis of the outward appearance characteristics of fruits and vegetables, thus enabling intelligent grading. In recent years, with the rapid advancement of deep learning technology, machine learning models based on neural networks have been extensively utilized in fruit and vegetable grading equipment. Deep learning models can learn and extract more complex and advanced features from large volumes of data, thereby endowing grading equipment with greater adaptability and intelligence.

On-site intelligent grading system for fruits and vegetables typically integrates sensor technology, image processing technology, and automation control technology. This enables automatic detection and evaluation of fruit and vegetable quality, size, shape, and other characteristics, thus achieving rapid and accurate grading and sorting of fruits and vegetables. With the continuous advancement of sensor technology, machine vision technology, and automation levels, intelligent grading equipment for fruit and vegetable origin has emerged as a significant research direction in agricultural engineering. It plays a crucial role in enhancing production efficiency, optimizing product quality, reducing waste, improving supply chain management, and enhancing market competitiveness. Consequently, it holds a pivotal role in driving modern agricultural production forward. Compared to traditional manual sorting and grading, On-site intelligent grading system for fruits and vegetables offers numerous advantages such as automation, efficiency, accuracy, data-driven operations, flexibility, and quality control. These advantages contribute to enhancing the modernization and competitiveness of fruit and vegetable production.

This paper provides a comprehensive review of the research progress in intelligent grading equipment for fruit and vegetable origin, aiming to explore the latest advancements in this field. Firstly, an overview of intelligent grading equipment for fruit and vegetable origin is provided, followed by a brief introduction to its structure. The paper primarily focuses on analyzing the application of machine vision technology in intelligent grading equipment for fruit and vegetable origin, along with research outcomes and advancements. It examines the current research status using tomatoes, apples, and other fruits and vegetables as examples. Analyze existing problems in the current development of this field and discuss future development trends.

4.2 Overview of On-Site Intelligent Grading System for Fruits and Vegetables

4.2.1 Basic Concepts of On-Site Intelligent Grading System for Fruits and Vegetables

On-site intelligent grading system for fruits and vegetables is a type of equipment that utilizes advanced technologies and devices to automate the detection and grading of fruits and vegetables. In modern agricultural production, with the continuous improvement of societal demands for food quality and safety, traditional manual sorting and grading methods have become insufficient to meet production requirements. Therefore, the introduction of intelligent technology and equipment has become an inevitable trend. The multi-faceted comparison between intelligent grading equipment and traditional manual grading is illustrated in Table 4.1. These devices typically integrate image recognition, sensor technology, and machine learning algorithms. Images of fruits and vegetables are captured through cameras or sensors, and computer vision algorithms are employed to analyze and process these images, enabling the identification of characteristics such as size, shape, color, and surface defects of the produce. Sensors are utilized to detect the physical attributes of the products, such as size, weight, hardness, and other relevant parameters.

Table 4.1 Comparison between intelligent grading equipment and traditional manual grading

Aspect	Intelligent grading equipment	Traditional manual grading
Productivity	High	Low
Cost	The initial cost is higher, but labor costs are saved in the long term	Labor costs are higher
Quality	Accurate grading to reduce errors	Affected by human factors
Production scale	Suitable for mass production	Suitable for small scale production
Data analysis	Real-time data analysis	Relying on manual experience makes quantitative analysis difficult

Commonly used sensors include laser sensors, pressure sensors, infrared sensors, among others, which are employed to collect real-time data of the products. Machine learning algorithms are capable of training on vast amounts of product sample data to learn the characteristics and quality standards of the products. Subsequently, they can be applied in practical scenarios for classification and grading. Furthermore, these algorithms can be continuously optimized and adjusted to enhance the speed and accuracy of grading. On-site intelligent grading system for fruits and vegetables from their origin can effectively enhance production efficiency, reduce labor costs, improve product quality, and ensure that fruits and vegetables meet market demands. Furthermore, the utilization of intelligent equipment can reduce waste and losses, increase the utilization rate of agricultural products, and promote the sustainable development of agricultural production. The emergence of On-site intelligent grading system for fruits and vegetables from their origin not only enhances the level of intelligence in agricultural production but also provides more development opportunities and competitive advantages for the grading of fruits and vegetables from their origin. With the continuous progress and application of technology, these intelligent equipment will receive wider application in the future, bringing more innovation and transformation to the development of the agricultural industry.

4.2.2 Structure and Components of On-Site Intelligent Grading System for Fruits and Vegetables

Intelligent grading equipment for fruits and vegetables from their origin is a device that integrates advanced technology with mechanical engineering. This paper introduces a common grading system, the main structure of which comprises feeding system, machine vision system, and sorting system. The structural diagram of On-site intelligent grading system for fruits and vegetables is illustrated in Fig. 4.2.

The fruit and vegetable feeding system is responsible for transporting harvested raw fruits and vegetables into the equipment. The machine vision system achieves

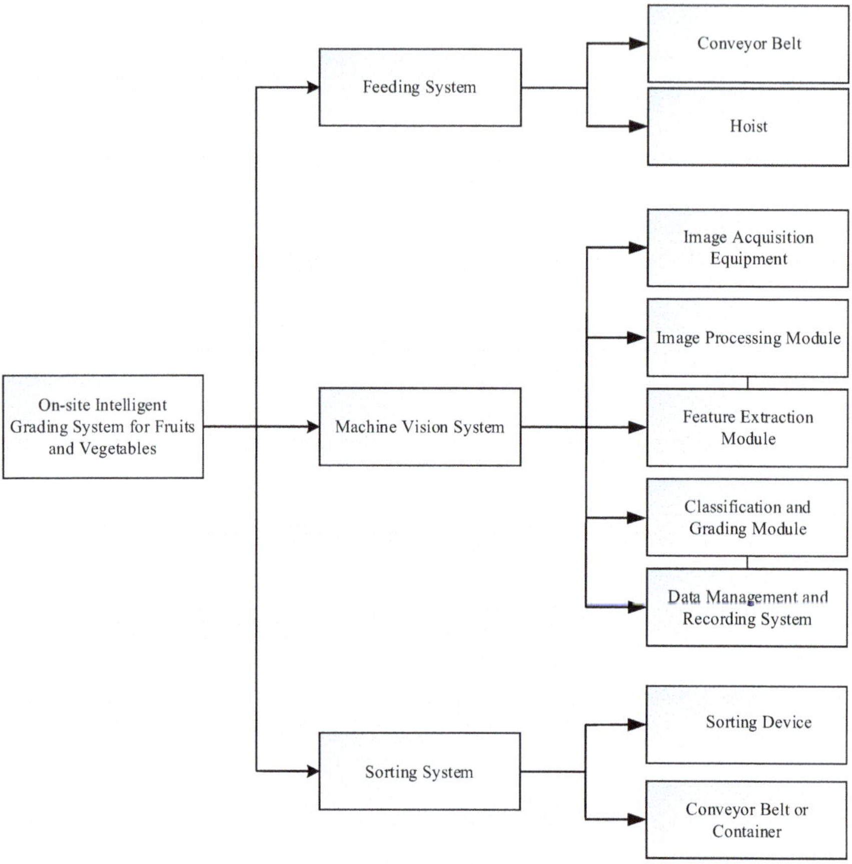

Fig. 4.2 Structural block diagram of on-site intelligent grading system for fruits and vegetables

rapid and accurate classification and grading of fruits and vegetables through automated image processing and analysis, thereby enhancing production efficiency and product quality.

The sorting system enables automated sorting and selection of fruits and vegetables, thus enhancing the level of automation and production efficiency of the production line.

(1) Feeding System

The feeding system is the part of the grading equipment that transports harvested raw fruits and vegetables. It mainly includes equipment such as conveyor belts, conveyors, and elevators, directly impacting production efficiency, product quality, and safety. As conveying equipment of the feeding system, the comparison between conveyor belts and elevators is illustrated in Table 4.2.

Through these automated conveying devices, fruits and vegetables can be transported directly from the original source to the grading equipment, reducing the costs

Table 4.2 Conveyor belt versus elevator comparison

Aspect	Conveyor belt	Hoist
Transportation method	Horizontal or tilted transport	Vertical conveying
Applications	Long distance, continuous operation	Scenarios with large ascent heights
Cost	Lower cost	Higher cost

and time associated with manual handling and transportation. This ensures the continuity and sustainability of the grading process. In 2016, Arakeri et al. utilized a conveyor belt as a feeding device and collected tomato images on the conveyor belt [10]. X. Zuo et al. invented a kiwifruit grading and conveying device that utilizes a feeding conveyor belt for kiwifruit loading. The device features a simple structure, significantly reducing the cost of kiwifruit grading [11]. From current research perspectives, conveyor belts as feeding device conveying equipment offer significant advantages. Their simple structure, suitability for continuous operation, and lower cost make them a preferred choice for many researchers. In contrast, the use of elevators is relatively less common, which is closely related to their specific application scenarios. When conveyor belts operate at high speeds or make turns, they may cause damage to delicate fruits and vegetables. Additionally, the level of automation of conveyor belts is not particularly high, leaving room for significant improvement. These are all issues worth considering in future research endeavors.

(2) Machine Vision System

On-site intelligent grading system for fruits and vegetables relies on the machine vision system. The machine vision system mainly consists of image acquisition devices, image processing modules, feature extraction modules, classification and grading modules, among other components. Image acquisition devices typically include industrial cameras, 3D cameras, etc. These devices often need to be used in conjunction with lighting sources to efficiently and effectively capture images of fruits and vegetables, ensuring consistency and stability in the captured images. The image processing module refers to the software or hardware components used to process, analyze, and recognize the captured images. The role of the feature extraction module is to extract information relevant to the characteristics of fruits and vegetables from the captured images, such as color, shape, texture, etc. The classification and grading module is the core component of the machine vision system, playing a crucial role in ensuring product quality, improving production efficiency, and reducing costs. During system operation, the conveyor belt is responsible for transporting fruits and vegetables, and the speed of the conveyor belt can be adjusted to control the movement speed of the produce. Once the sensors detect that the fruits and vegetables have reached their positions, they send a triggering signal to the controller, indicating that they are ready for image acquisition. Upon receiving the triggering signal from the sensors, the controller activates corresponding operational steps based on preset programs and algorithms. These steps include triggering the industrial camera for image acquisition and controlling the lighting source to

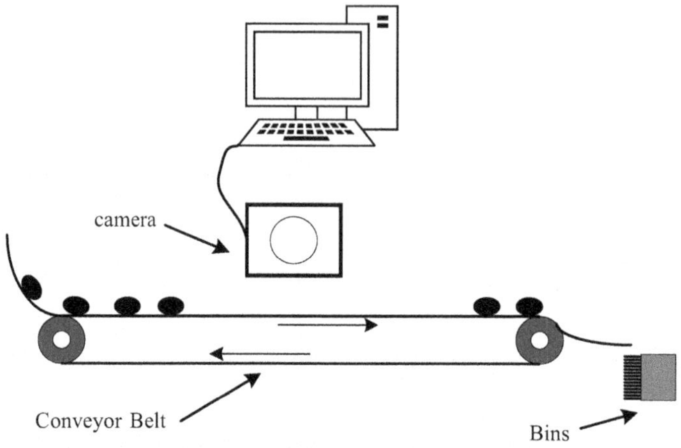

Fig. 4.3 [10] System structure diagram

provide appropriate illumination. After receiving the images captured by the industrial camera, the controller continues to perform processes such as image processing, feature extraction and analysis, as well as decision-making and control. The system architecture diagram in [10] is shown in Fig. 4.3.

(3) Sorting System

The sorting system classifies and sorts fruits and vegetables based on the information provided by the machine vision system. The sorting system mainly consists of sorting devices, control modules, and conveyors or containers. Sorting devices are used to actually classify and sort fruits and vegetables, commonly used sorting devices include pickers, pushers, etc., which are advantageous due to their simple structure and low cost. Wang Lijuan et al. developed a color sorting machine for apples, employing pushers as sorting devices. By driving the motor, the pushers are moved by extension rods to push the fruits out of the conveyor belt and into the collection container [12]. The push plate structure in [12] is shown in Fig. 4.4.

The selection of sorting devices needs to consider minimizing damage to the fruits while ensuring that the working speed and accuracy meet the actual production requirements. Additionally, factors such as equipment cost and spatial layout need to be taken into account. Conveyors or containers serve as the connecting devices between the feeding system and the sorting device. They are responsible for transporting fruits and vegetables to be sorted and delivering sorted fruits and vegetables to the next processing stage.

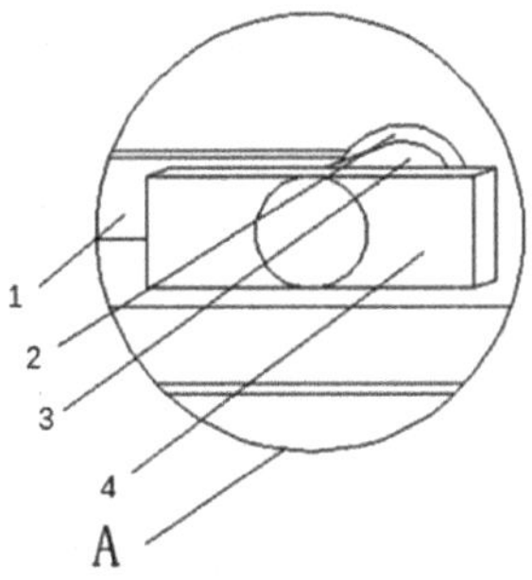

1. Connection board 2. Push motor
3. Telescopic pole 4. Push the board

Fig. 4.4 Push plate structure diagram in [12]. 1. Connection board 2. Push motor 3. Telescopic pole 4. Push the board

4.3 Application of Machine Vision Technology in On-Site Intelligent Grading System for Fruits and Vegetables

Machine vision technology utilizes devices such as cameras to obtain images of fruits and vegetables, and through image processing and analysis algorithms, it achieves the recognition and analysis of features such as size, shape, and color. This enables automated classification and grading of fruits and vegetables, enhancing production efficiency and product quality.

4.3.1 Overview of Machine Vision Technology

Machine vision technology is a technique that simulates and implements the capabilities of the human visual system using computer systems, extracting information, analyzing, and processing data from images or video data. It utilizes cameras or other imaging devices to capture image data, then processes the images using algorithms such as image processing, pattern recognition, and deep learning to analyze and understand the images, thereby accomplishing automated visual tasks. The application of machine vision technology is becoming increasingly widespread in On-site intelligent grading system for fruits and vegetables. Through machine vision technology, equipment can acquire information about the appearance, size, color, defects, and other features of fruits and vegetables, enabling rapid and accurate detection and classification of the produce. Specifically, machine vision technology can assist equipment in achieving functions as shown in Table 4.3.

With the development and application of machine vision technology, On-site intelligent grading system for fruits and vegetables can achieve efficient and precise grading, thereby enhancing production efficiency, reducing costs, ensuring product quality, and meeting market demands. The integration of this technology not only

Table 4.3 The function of machine vision technology in intelligent grading equipment in fruit and vegetable producing areas

Function	Describe
Image acquisition	Use imaging equipment to capture images or video data of fruits and vegetables
Feature extraction	Extract characteristic information of fruits and vegetables, such as size, shape, color, etc
Pattern recognition	Identify different grades or classifications of fruits and vegetables
Defect detection	Detecting blemishes or defects on the surface of fruits and vegetables
Data analysis	Analyze and process the collected data

enhances the production level of fruits and vegetables from their origin but also brings more opportunities and development space for the agricultural industry.

4.3.2 Basic Principles of Machine Vision

The working principle of machine vision technology involves simulating the human visual system to process, analyze, and recognize image or video data using computer systems.

In On-site intelligent grading system for fruits and vegetables, the working principle flowchart of machine vision technology is illustrated in Fig. 4.5.

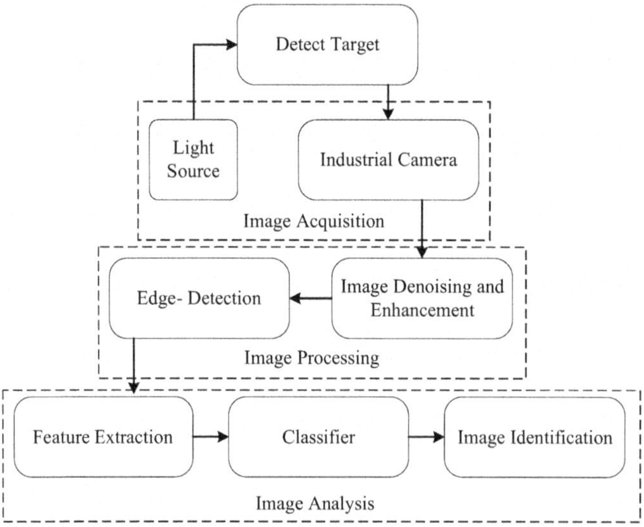

Fig. 4.5 Working principle flow chart of machine vision technology

First, the fruits and vegetables are photographed using imaging equipment to obtain image data. After image acquisition, preprocessing is performed on the images, including noise removal and adjustments to brightness and contrast, to enhance the accuracy of subsequent analysis. After image preprocessing is completed, the next step is to extract feature information from the images, such as size, shape, color, etc., of the fruits and vegetables. Next, the extracted features are compared with predefined standard features through feature matching to determine whether the fruits and vegetables meet specific standards or grades. Subsequently, pattern recognition algorithms are employed to classify and identify the features of fruits and vegetables, determining their respective grades or categories. Finally, based on the recognition results, the intelligent grading equipment makes corresponding decisions, such as allocating fruits and vegetables to appropriate grades or categories, and carries out automatic sorting and packaging through mechanical execution systems. The aforementioned steps constitute the workflow of machine vision technology in On-site intelligent grading system for fruits and vegetables. Through these processes, the equipment can achieve automated grading of fruits and vegetables, thereby improving production efficiency and product quality. The following sections will provide a brief introduction to processes such as image acquisition, image preprocessing, and external feature extraction.

4.3.3 Image Acquisition

Image acquisition refers to the process of obtaining fruit and vegetable image data using devices such as cameras or sensors. It is the first and crucial step in intelligent grading equipment because subsequent tasks such as fruit and vegetable grading and recognition are based on the collected image data. Image acquisition needs to meet requirements such as real-time performance, stability, high definition, and multi-angle acquisition to ensure the acquisition of accurate and reliable image data, providing effective support for subsequent intelligent grading work. Lighting is crucial for the image acquisition process. The stability, brightness, and color temperature of the light source directly affect the clarity, contrast, and color accuracy of the images. An appropriate light source can provide sufficient brightness and uniformity, ensuring clear contours and rich details of objects in the images, thereby improving the accuracy of subsequent image processing and analysis. Common types of machine vision light sources are illustrated in Fig. 4.6. The characteristics of strip lights include their compact size and adaptability to different positions, primarily used for detecting larger rectangular objects. Ring lights can be used to detect scratches and damages on object surfaces. Backlights can be utilized for detecting edge defects and surface flatness, among other purposes.

In 2009, LI et al. used a color CCD camera to capture images and output color information from the red, green, and blue channels. However, at that time, CCD cameras were very expensive [13]. Kumar et al. automated mango grading based on machine vision technology, utilizing a Raspberry Pi V2 camera for image acquisition,

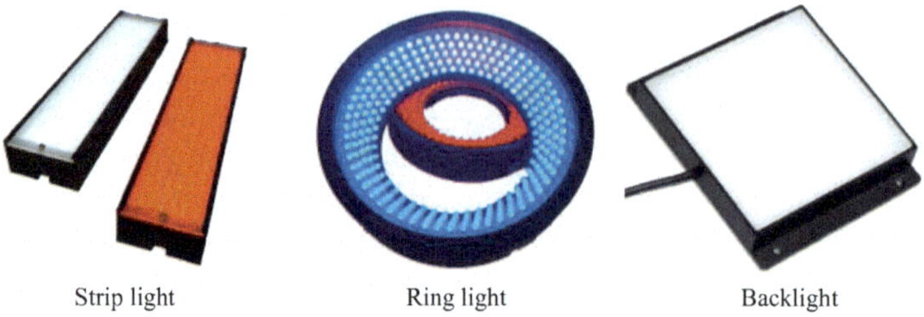

Fig. 4.6 Common light source types

which was relatively inexpensive in price [14]. Das et al. utilized a Nikon D3500 CCD camera to capture tomato images in the field, creating a dataset that yielded high-quality images. However, the price of this camera is relatively expensive [15]. Under the guidance of domain experts, Khatun et al. used Redmi 11 Pro Plus and Samsung S22 cameras to capture 800 × 800 pixel images of dragon fruits, storing them in JPG format for subsequent dragon fruit grading work [16]. Jaramillo-Hernandez et al. utilized digital cameras equipped with photoelectric sensors to acquire 2D RGB images of fruits and vegetables, which were subsequently used for tasks such as detecting fruit size and color. An advantage of this approach is its relatively low cost [17]. In summary, researchers usually opt for more expensive and superior-performance cameras to capture higher-quality images during fruit and vegetable image acquisition, aiming to ensure better image quality.

4.3.4 Image Processing and Analysis Techniques

Image processing and analysis techniques refer to the technology that utilizes computer and digital image processing methods to process, analyze, and understand images.

It involves various operations and algorithmic processing of images to extract information, features, or achieve specific objectives from the images. Common image processing and analysis techniques include image denoising and enhancement, edge detection, and image segmentation.

4.3.4.1 Image Preprocessing

Image preprocessing refers to a series of processing steps applied to an image before further image processing. These steps aim to improve the quality of the image, reduce noise, enhance image features, eliminate interference, etc., in order to better adapt to subsequent image processing tasks. Image denoising techniques mainly include mean filtering, median filtering, and Gaussian filtering. The comparison of these three methods is shown in Table 4.4.

Table 4.4 Comparison of denoising methods

Feature	Mean filter	Median filter	Gaussian filter
Principle	Replace each pixel in the image with the average of its surrounding pixels	Replace each pixel in the image with the median of its surrounding pixels	Filter the image using Gaussian function
Applicability	Simple tasks that require quick noise removal	Suitable for removing outliers such as salt and pepper noise	Tasks that require image detail preservation
Effect	Has some effect on mild noise, but may result in loss of image details	It has a good suppression effect on various types of noise	Better effect on preserving image details

Image preprocessing plays a crucial role in the application of machine vision technology in On-site intelligent grading system for fruits and vegetables. On-site intelligent grading system for fruits and vegetables requires accurate classification of the produce. Image preprocessing helps improve image quality by eliminating noise and interference, thereby enhancing the accuracy and stability of classification. During grading, it is necessary to extract and recognize features of fruits and vegetables. Image preprocessing can help highlight key features of the produce, thereby improving their recognizability and recognition accuracy. On-site intelligent grading system for fruits and vegetables needs to operate under complex environmental conditions, such as uneven lighting and foreign objects on the surface of the produce. Image preprocessing can help mitigate the impact of environmental conditions on image acquisition, thereby improving the recognition stability and robustness of the produce. Sawant et al. used morphological filters to improve image contrast and eliminate noise [18]. In the fruit grading system developed by Pushpavalli, a median filter was employed to suppress noise present in the captured images [19]. Usha et al. compared the denoising effects of Wiener filtering, adaptive filtering, and median filtering, finding that median filtering yielded superior results. Therefore, they opted for median filtering to remove noise and improve image quality [20]. Doan et al. employed various data augmentation techniques such as horizontal flipping and rotation to enhance the original dataset, thereby improving the model's generalization ability and preventing overfitting [21]. Median filtering is widely used in image denoising because it is a highly effective denoising method. However, in some cases, it may lead to the blurring or smoothing of image details. Moreover, median filtering may not be effective in handling Gaussian noise. Therefore, when selecting a filter, comprehensive considerations should be made.

4.3.4.2 Image Processing

The primary goal of image processing is to perform specific operations on images to obtain new images or extract useful information. Image processing aims to alter the appearance, content, or structure of images to meet specific requirements, such as image enhancement, edge detection, segmentation, etc.

The main methods of image enhancement include histogram equalization, contrast stretching, and sharpening. Histogram equalization is a method of adjusting the distribution of pixel values in an image to make it more uniform. Contrast stretching is also a simple image enhancement method that enhances contrast by linearly mapping the pixel values of an image to a wider range. Sharpening enhances the edge information of an image, making it appear clearer. Kumari N et al. employed a median filter to remove noise from the image and used image morphology functions to enhance image contrast [22]. Deepa et al. compared the performance of contrast adjustment, noise removal, histogram equalization, and binarization in mango image preprocessing, and evaluated the performance of these methods using peak signal-to-noise ratio (PSNR) and mean square error (MSE) [23].

Edge detection aims to identify boundaries or regions between objects in an image.

Common edge detection algorithms include Canny edge detection, Sobel operator, and Prewitt operator. Bautista et al. applied image processing techniques to analyze the quality of tomatoes, accurately detecting edges using the Canny edge detection algorithm [24]. Zárate and colleagues applied the Canny edge detection algorithm to locate regions of interest (ROI) within the image, and then extracted features of the fruits [25]. In [26], tomato regions of interest were detected, as shown in Fig. 4.7.

Hussin and colleagues utilized edge detection algorithms such as Canny, Laplacian, and Sobel for processing durian images. Experimental results indicated that the Laplacian edge detection algorithm performed the best [27]. In summary, the application of the Canny edge detection algorithm is more extensive. It is a highly accurate and robust edge detection algorithm, with broad prospects for application in the field of fruit and vegetable grading.

Image segmentation is the process of accurately separating fruit and vegetable objects from the background in an image. Typically, image segmentation is used to extract the contours or boundaries of fruits and vegetables for tasks such as shape analysis, quality assessment, classification, and grading. Commonly used image segmentation methods include threshold segmentation, clustering segmentation, and deep learning-based segmentation. Kaur et al. used the Otsu method in their developed

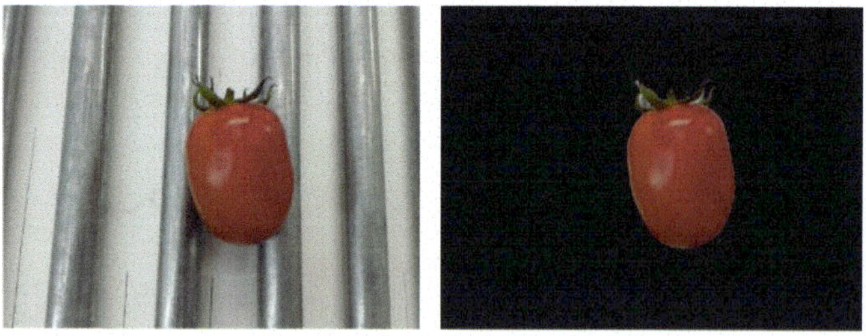

original image region-of-interest

Fig. 4.7 original image and region of interest of tomato in [26]

automatic tomato grading and sorting system to segment and separate vegetables from the background [28].

Pawgi et al. utilized the k-means algorithm to segment the captured fruit images [29]. Waqar et al. employed threshold segmentation for image segmentation, initially converting the image to a grayscale image, and then converting the grayscale image into a binary image [30]. Vite-Chavez et al. improved the segmentation of fruit and vegetable images based on the OTSU method, which can overcome uneven lighting and other issues, achieving the best segmentation results [31]. The OTSU method is widely chosen by many researchers for fruit and vegetable image segmentation, making its application quite extensive in fruit and vegetable grading. This method is simple, efficient, easy to implement, and computationally fast, resulting in good performance in practical applications.

4.3.4.3 External Feature Extraction

External feature extraction is also a crucial step in vegetable grading. In the research of On-site intelligent grading system for fruits and vegetables, external feature extraction refers to the process of extracting important information or features used for grading and identification from the external characteristics of vegetables' surfaces. These external features can include the shape, color, size, surface texture, and other characteristics of the vegetables. By extracting these features, intelligent grading equipment can accurately identify and grade different types of vegetables. Color features are crucial for identifying the ripeness and quality of vegetables. They can be described using methods such as histogram statistics and color space conversion (such as RGB, HSV, etc.) to depict the distribution of colors in different parts. The images of Apple under RGB and HSV models are shown in Fig. 4.8 [32].

Analyzing the texture features of vegetable surfaces, such as wrinkles and unevenness, can help determine their freshness and quality. Common texture feature extraction methods include Gray-Level Co-occurrence Matrix (GLCM) and Local Binary Patterns (LBP). PCA (Principal Component Analysis) provides an effective method for feature extraction. It can identify the feature vectors that best represent the variability in the original data, projecting the raw data onto these feature vectors to obtain a more compact and efficient feature representation. Based on deep learning, feature extraction typically utilizes Convolutional Neural Networks (CNN) as the primary tool. CNN is a type of deep learning model specifically designed to handle image data, capable of automatically learning feature representations within images. By extracting the external features of vegetables and fruits, intelligent grading equipment can more accurately identify their types, quality, and ripeness, enabling automated grading and classification of produce. Olaniyi et al. used Gray-Level Co-occurrence Matrix (GLCM) for feature extraction to classify mango ripeness [33]. Bhargava and collaborators employed Principal Component Analysis (PCA) to extract external features of apple fruits, providing support for subsequent category grading [34]. Kumar and Sudha, and their team proposed a method that utilizes Histogram of Oriented Gradients (HOG) and Local Binary Patterns (LBP) to extract external features of pomegranates, enhancing the accuracy and efficiency of grading [35]. As of current research, there are numerous methods for extracting external features from images, and researchers strive to achieve the best feature extraction results by using different algorithms.

4.3.5 Classification and Grading Algorithms

On-site intelligent grading system for fruits and vegetables uses grading and classification algorithms to automatically distinguish and classify fruits and vegetables.

Selecting an appropriate classification algorithm can significantly enhance the efficiency of classification and sorting, thereby reducing the time and cost of manual sorting, while ensuring the quality of sorting. Commonly used classification algorithms for fruit and vegetable grading include Support Vector Machine (SVM), Random Forest (RF), and K-Nearest Neighbors (KNN), among others.

Support Vector Machine (SVM) is a classic classification algorithm commonly employed in fruit and vegetable grading tasks for binary or multiclass classification of fruit and vegetable images. It performs well on smaller datasets and exhibits good generalization capabilities for high-dimensional data. Mapping the feature vectors of instances (using two dimensions as an example) to points in space, as illustrated in Fig. 4.9.There are two different types of points, and the objective of SVM is to draw a line that "best" separates these two types of points, such that if new points are introduced later, this line can also classify them effectively.

Jiewen, Z et al. proposed a method based on Support Vector Machine (SVM) for identifying defects and stems in Fuji apples. Experimental results demonstrated an accuracy of 90% [36]. Azarmdel et al. developed a machine vision-based system for mulberry ripeness detection, which employs Artificial Neural Networks (ANN) and Support Vector Machine (SVM) for mulberry classification [37]. Patel et al. proposed a novel integrated approach, the Stacked Sparse Denoising Autoencoder-Support Vector Machine (SSDAE-SVM) method, for fruit grading and classification. This method simplifies fruit grading and alleviates post-harvest losses [38]. In conclusion,

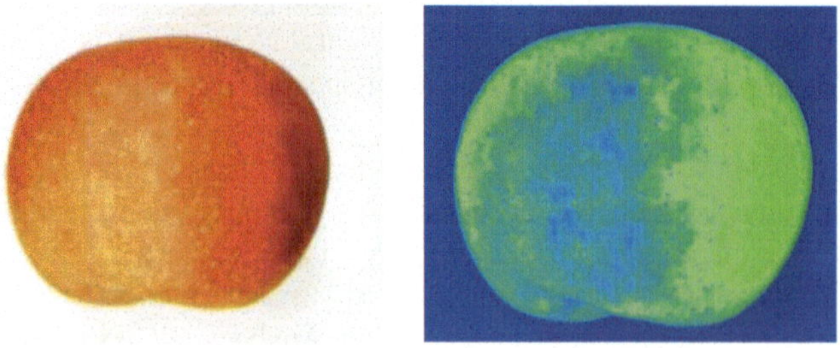

RGB model HSV model

Fig. 4.8 Apple images in RGB and HSV models

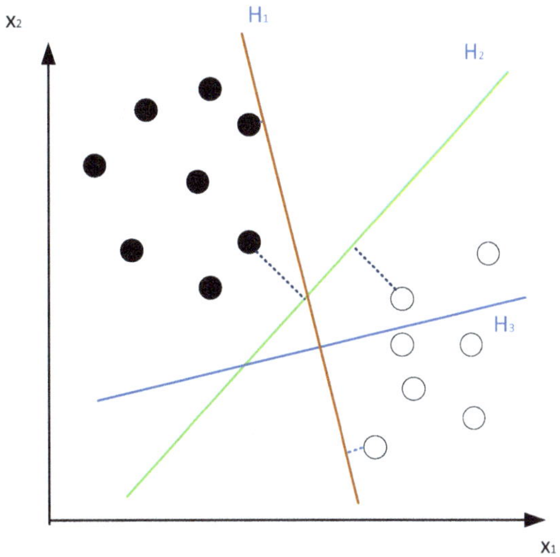

Fig. 4.9 SVM concept diagram

Support Vector Machine (SVM) stands as a classic and powerful machine learning algorithm that has been applied early on in fruit and vegetable grading tasks and continues to be utilized to this day.

The Random Forest (RF) algorithm is a machine learning method based on ensemble learning, commonly employed for classification and regression tasks. The basic principle of the RF algorithm involves constructing multiple decision trees and aggregating the prediction results of each tree through methods such as voting or averaging. The construction process of each tree is based on randomly selected training samples and features. This randomness aids in reducing overfitting and improving the model's generalization capability. The principle diagram of Random Forest is illustrated in Fig. 4.10.

In fruit and vegetable grading tasks, the RF algorithm can be applied to automate the classification and grading of fruits and vegetables. Tripathi et al. employed ensemble classifiers for mango grading, where the classifier incorporates the Random Forest (RF) algorithm [39]. In mango grading tasks, Naik et al. utilized Support Vector Machine (SVM) and Random Forest (RF) for classification and conducted comparisons [40]. The RF algorithm exhibits high accuracy and robustness in fruit and vegetable grading, making it a commonly used and effective classification algorithm.

The K-Nearest Neighbors (KNN) algorithm is a type of instance-based supervised learning algorithm. The algorithm computes the distance between a new sample and samples in the training set, then selects the K nearest neighbors based on this distance. It classifies or predicts the new sample based on the labels of these neighbors. In their developed eggplant grading system, Akter et al. utilized the KNN algorithm to classify eggplants into four categories: healthy, partially defective, moderately defective, and unhealthy, achieving a success rate of 88% [41]. Ayuningtyas et al. utilized K-Nearest Neighbors (KNN) and Support Vector Machine (SVM) for tomato

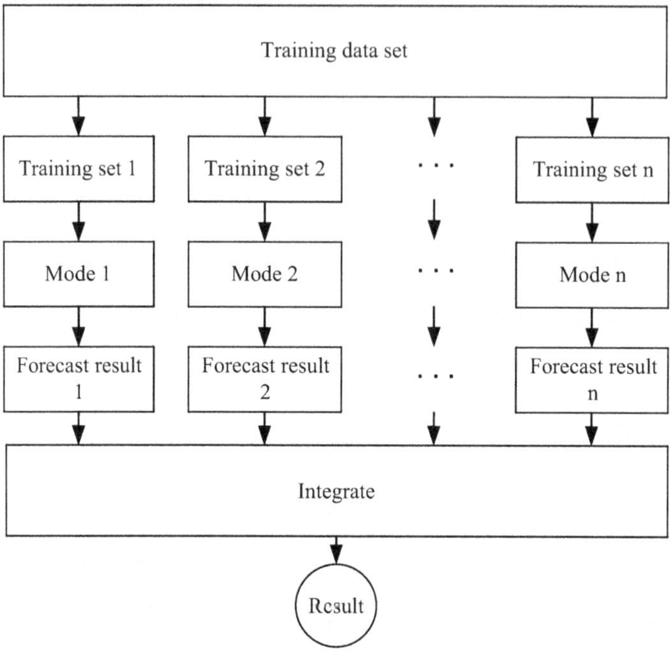

Fig. 4.10 Random forest schematic diagram

ripeness classification. Experimental results demonstrated that the accuracy of the KNN method surpassed that of SVM [42]. Trieu et al. employed K-Nearest Neighbors (KNN) for classifying whether dragon fruit skin is defective. Following experimentation, the classification accuracy reached 98.5% [43]. KNN is also a traditional machine learning algorithm known for its simplicity and ease of understanding and implementation. It is widely applied in fruit and vegetable grading tasks due to its straightforward concept.

4.4 Existing Research Achievements of Machine Vision Technology in the Field of Fruit and Vegetable Grading

With the rapid development of artificial intelligence technologies such as computer vision and deep learning, machine vision has made significant progress in image recognition, object detection, and other areas. This progress has provided technical support for the realization of automatic grading in the field of fruit and vegetable intelligent equipment. In order to meet the demands for automation, utilize technological advancements, enhance quality control, and improve production efficiency, extensive research has been conducted on non-destructive rapid detection through machine vision.

In 1990, Shearer et al. utilized machine vision technology to grade agricultural products based on color. Their method achieved a classification accuracy of 96% for bell pepper quality grading [44]. In 1996, Engle et al. quantitatively defined key parameters for grading and sorting tomatoes and fresh honeydew melons. They developed a neural network-based model for agricultural product quality sorting. This research enabled non-destructive testing and yielded consistent results with manual sorting by operators [45]. From the above, it can be seen that research on intelligent grading methods for fruits and vegetables started early. These advanced technologies can be applied in intelligent equipment for fruit and vegetable production areas, continuously leading the fruit and vegetable industry towards intelligent and efficient development, injecting new vitality into the improvement and efficiency of the fruit and vegetable industry.

4.4.1 Status of Tomato Grading

Tomatoes contain abundant nutrients such as vitamin C, vitamin A, and potassium, which contribute to maintaining the health and functionality of the body. Tomatoes are primarily graded based on parameters such as color, shape and size, ripeness, and presence of defects. Referring to the Chinese industry standard GH/T 1193—2021 [46], the classification criteria for tomatoes are as shown in Table 4.5.

Currently, research on machine vision technology for tomato grading has made significant progress, with many scholars and engineers dedicated to developing efficient and accurate automated grading systems. Muhaemin proposed an automated tomato grading machine based on visual assessment and developed an image processing program for tomato classification. The accuracy can reach 95.5%, and it can grade 1200 tomatoes per hour. However, during the grading process, 4% of the tomatoes may incur mechanical bruising [47]. Arakeri et al. proposed a method for

Table 4.5 GH/T 1193—2021 tomato classification standard

Grade	Requirements
Special Grade	Consistent appearance, good fruit shape and color, smooth fruit skin, fresh green calyx; firm and elastic fruit, appropriate ripeness; free from sunburn, discoloration, scars, hail damage, frost damage, wrinkling, and mechanical injuries
Grade A	Appearance is generally consistent, with good fruit shape and color, relatively smooth fruit skin, and relatively fresh green calyx; the fruit is firm and elastic, either ripe or slightly underripe; free from sunburn, discoloration, scars, hail damage, frost damage, wrinkling, misshapen fruit, cracked fruit, and mechanical injuries
Grade B	Appearance is generally consistent, with reasonably good fruit shape and color, and slight wilting of the calyx; the fruit is moderately firm, with slightly reduced elasticity, and is either slightly underripe or slightly overripe; free from severe sunburn, misshapen fruit, cracked fruit, and mechanical injuries, with no significant impact on fruit marketability

automatic tomato grading using computer vision technology, which classifies tomatoes into categories such as defective, non-defective, ripe, and unripe. The accuracy rates for defective/non-defective and ripe/unripe classification are 100% and 96.47%, respectively. However, the system can only grade 300 tomatoes per hour, indicating significant room for improvement in grading speed [10]. Lu et al. proposed a tomato grading system based on DenseNet-201. The system utilizes transfer learning and data augmentation techniques to generate an augmented training set for model training, achieving high classification accuracy [48] Opeña et al. proposed an automatic tomato grading system utilizing an Artificial Bee Colony (ABC) algorithm for training and an Artificial Neural Network (ANN) classifier. This system effectively classifies the ripeness of tomatoes [49]. Liu et al. developed an intelligent tomato grading system. This system selects color, shape, and size as key features and establishes an automatic grading classifier based on tomato surface characteristics. They also constructed a grading platform to validate the classifier's grading performance. The average grading accuracy of this method is 90.7%. However, the research faces the challenge of relatively slow grading speed [50]. Deulkar et al. utilized a digital camera to capture tomato images, employed the Otsu thresholding and K-Means clustering algorithms to extract fruit features, and utilized Support Vector Machine (SVM) to achieve fruit quality sorting, thus realizing an efficient automatic non-destructive grading system [51]. Yusiong proposed a CNN-ELM model for tomato ripeness grading. This model achieves a classification accuracy of 96.67% with a limited dataset [52]. Li et al. proposed a tomato ripeness grading and counting model based on MHSA-YOLOv8. This model supports both online monitoring and offline detection, greatly assisting in the efficiency of tomato grading at production sites [53].

Mputu et al. proposed a tomato classification and grading method that utilizes pre-trained Convolutional Neural Networks (CNN) for feature extraction and traditional machine learning algorithms (SVM, RF, KNN) for classification. Through experimental comparisons, the CNN-SVM hybrid model performed the best in tomato grading, achieving an accuracy of 96.67%. However, this system cannot perform size grading, and overfitting may occur when the training dataset is too small [26]. Dao et al. proposed a deep learning network model based on YOLOv8 for detecting tomato surface defects. They trained the dataset using 500 tomato images, achieving an accuracy of 96.3%. This demonstrates the effective application of YOLOv8 in tomato quality inspection systems [54]. Appe et al. proposed a method for tomato ripeness detection and classification based on an improved YOLOv5 tomato detection algorithm. This method utilizes YOLOv5 for feature extraction and object recognition, and incorporates Convolutional Block Attention Module (CBAM) to enhance accuracy. Experimental results show that this method achieves higher efficiency in detecting smaller tomatoes, with an average precision of 88.1% [55]. Huang et al. utilized fluorescence imaging and an improved YOLOv5m model to detect multiple defects in tomatoes. The accuracy of the improved model reached 87.57%, surpassing YOLOv5m, Faster R-CNN, and YOLOv7 [56]. Amna and others designed an automatic fruit grading system based on machine vision technology. The system also has a mechanical sorting device that can sort the classified fruits. Through grading tests on mangoes and tomatoes, the accuracy rates reached 95% and 94% respectively [57].

4.4.2 Status of Tomato Grading

The classification criteria for apples, as a common type of fruit, typically include parameters such as shape, size, color, and surface defects. Huang et al. proposed an apple defect region identification method based on Gabor wavelet transform and machine vision support vector machine (SVM). Experimental results indicate an identification accuracy of 85% [58]. Yuan et al. proposed a grading method based on an improved Particle Swarm Optimization (PSO) algorithm and Support Vector Machine (SVM). This method utilizes an intelligent PSO algorithm to select the optimal parameters for the kernel function in SVM, enhancing the classifier's performance. Ultimately, the trained SVM is employed to grade apple color [59].

Bhatt et al. proposed an apple classification system based on machine vision and Artificial Neural Networks (ANN), capable of real-time classification of apple size, color, and external defects parameters [60]. Gongal et al. developed a machine vision system for estimating apple fruit size, consisting of a color CCD camera and a Time-of-Flight (TOF) light-based 3D camera. Through experimental analysis, the accuracy of apple size estimation using pixel size was found to be 84.8%, while using the 3D coordinate method yielded an accuracy of 69.1%. Clearly, estimating apple size based on pixel size achieves higher accuracy. The error in pixel size estimation is attributed to inaccurate image segmentation, whereas the error in 3D coordinate estimation primarily stems from the low resolution of the 3D sensor. Future improvements could involve enhancing image processing methods to achieve more accurate grading [61]. Bhargava et al. developed an apple quality grading system utilizing fuzzy c-means clustering for defect region segmentation, followed by classification using an SVM classifier. The system achieved an accuracy of 98.48% [62]. Liang et al. proposed a semantic segmentation method based on the BiSeNet V2 deep learning network for segmenting apple defect regions. They further enhanced the detection accuracy of defect areas in apple images using an improved YOLOv4 network. Experimental results demonstrated an accuracy of 92.42% [63]. Zhao et al. employed an enhanced Grey Wolf Optimizer (GWO) to optimize the penalty and kernel parameters within the Support Vector Machine (SVM) framework, obtaining the optimal IGWO-SVM classification model for external quality assessment of apples. The accuracy of this model surpasses that of the GMO-SVM classification results [64]. Ji et al. introduced a multi-attribute information processing method for apple grading based on the YOLOv5s network architecture. This method enables rapid and accurate completion of apple quality grading tasks, achieving an accuracy of 94.46% [65]. Xu et al. proposed an enhanced YOLOv5 apple grading method. Experimental results demonstrate an apple grading accuracy of 90.6%, with a grading speed of four apples per second [66]. Iosif et al. optimized several CNN models and found that MobileNet and Inception models exhibited the highest grading accuracy.

They underscored the potential of machine learning algorithms in enhancing apple quality grading. [67].

4.5 Conclusion and Outlook

Currently, research on On-site intelligent grading system for fruits and vegetables has achieved fruitful results but still faces a series of challenges and issues. For example, there are issues related to accuracy and stability in the technology aspect, and achieving a balance between accuracy and speed in fruit and vegetable grading is quite challenging. Although there are already some intelligent grading equipment available, their accuracy and stability still need improvement. Cost is also a constraining factor, as high research and development costs and equipment procurement expenses limit the widespread adoption of intelligent grading equipment. Different types of fruits and vegetables require targeted grading equipment, which increases the complexity of research and production. The diversity in the characteristics of agricultural products, such as shape, size, and color, also adds to the difficulty of fruit and vegetable grading. How to comprehensively consider these factors is a challenging issue worth pondering.

In the future, intelligent grading equipment will evolve towards being more efficient, accurate, and intelligent. With the continuous progress and development of artificial intelligence, intelligent grading equipment will become more intelligent, capable of automatically adjusting grading parameters based on the characteristics of fruits and vegetables to enhance accuracy and stability. The upgrading of sensors and the development of image recognition technology will further improve the perception capability of grading equipment, enabling better recognition of the appearance characteristics of fruits and vegetables and achieving more precise grading. Additionally, advancements in manufacturing technology will gradually reduce the cost of intelligent grading equipment, making fruit and vegetable grading tasks more efficient and of higher quality. Intelligent grading equipment may also integrate with Internet of Things (IoT) technology, enabling interconnection between devices to improve production efficiency and data analysis capabilities. Overall, in the future, intelligent grading equipment will become one of the essential equipment for fruit and vegetable production areas, driving industry upgrades and enhancing competitiveness.

References

1. Akter T et al (2024) A comprehensive review of external quality measurements of fruits and vegetables using nondestructive sensing technologies. J Agricult Food Res 15
2. FAO, Agricultural production statistics 2000–2020. 2022, Rome, Italy. ISSN 2709–0078
3. Nicolaï BM et al (2014) Nondestructive measurement of fruit and vegetable quality. Annu Rev Food Sci Technol 5:285–312

4. Baswaraj D et al (2023) Importance of machine vision framework with nondestructive approach for fruit classification and grading: a review. Int J Recent Innovat Trends Comput Commun
5. Jaydeep B, Patil SBP (2023) Deep learning for fruit grading: a state-of-the-art review. Int J Recent Innovat Trends Comput Commun
6. Ratha AK et al (2023) Computer intelligence-based fruit grading: a review. Revue d'Intelligence Artificielle 37(2):465–474
7. AmeethaJunaina Mrs TK et al (2020) A survey on fresh produce grading algorithms using machine learning and image processing techniques. IOP conference series: materials science and engineering 2020:981
8. Pokhrel B (2021) Review on post-harvest handling to reduce loss of fruits and vegetables. Int J Hortic Food Sci 2:48–52
9. Daheriya M et al (2023) Advances in multi-fruit and vegetable grading: a comprehensive review. Int J Environ Clim Change 13(9):1239–1245
10. Arakeri MP, Lakshmana (2016) Computer vision based fruit grading system for quality evaluation of tomato in agriculture industry. Proc Comput Sci 79:426–433
11. Zuo X, Xu L (2013) Design of control system for kiwifruit automatic grading machine. Sens Transduc 21(5):58
12. Wang L et al (2022) Design of apple color sorting machine for machine vision maturity detection. Agricult Technol 42(12):36–40
13. Li C, Cao Q, Guo F (2009) A method for color classification of fruits based on machine vision. WSEAS Trans Syst Arch 8:312–321
14. Kumar AGS et al (2019) Machine vision technique based smart fruit sorter. IRE J
15. Das P, Yadav JKPS, Singh L (2022) Deep learning-based tomato's ripe and unripe classification system. Int J Softw Innov 10:1–20
16. Khatun T et al (2024) A comprehensive dragon fruit image dataset for detecting the maturity and quality grading of dragon fruit. Data Brief 52:109936
17. Jaramillo-Hernandez JF et al (2024) Application of machine vision techniques in low-cost devices to improve efficiency in precision farming. Sensors (Basel) 24(3)
18. Sawant AA, Kshirsagar V (2014) Evaluation of segmentation algorithms for apple fruit grading
19. Pushpavalli M (2019) Image processing technique for fruit grading. Int J Engin Adv Technol
20. Usha S et al (2017) Automated sorting and grading of vegetables using image processing. Int J Engin Res Gener Sci 5(6):53–61
21. Doan T-N, Le-Thi D-N (2023) A novel mango grading system based on image processing and machine learning methods. Int J Adv Comput Sci Appl
22. Kumari N et al (2020) Automated grading of mangoes based on surface defect detection using a combined approach of image segmentation. Environ Conservat J
23. Deepa PV, Geethalakshmi DSN (2011) Comparison of image preprocessing techniques for fruit grading. Digit Image Process 3:824–828
24. Bautista JF et al (2020) Automated sorter and grading of tomatoes using image analysis and deep learning techniques. In: 2020 IEEE 12th international conference on humanoid, nanotechnology, information technology, communication and control, environment, and management (HNICEM), pp 1–6
25. Zárate V, González E, Cáceres-Hernández D (2023) Fruit detection and classification using computer vision and machine learning techniques. In: 2023 IEEE 32nd international symposium on industrial electronics (ISIE). IEEE
26. Mputu HS et al (2024) Tomato quality classification based on transfer learning feature extraction and machine learning algorithm classifiers. IEEE Access 2024(12):8283–8295
27. Hussin AAA et al (2023) Utilizing different edge detection and preprocessing techniques to improve the accuracy of durian cultivar detection using convolutional neural networks. In: 2023 IEEE 9th international conference on computing, engineering and design (ICCED), pp 1–6
28. Kaur S, Girdhar A, Gill J (2018) Computer vision-based tomato grading and sorting
29. Pawgi A et al (2022) Automated fruit quality detection using image processing. J Image Process Intell Remote Sens

30. Waqar M et al (2023) Machine vision-based automatic fruit quality detection and grading. Front Agricult Sci Engin
31. Vite-Chavez O et al (2023) Improvement procedure for image segmentation of fruits and vegetables based on the Otsu method. Image Anal Stereol
32. Liu JiaHao et al (2023) Fruit grading system based on machine vision. Food Mach 39(06):112–118
33. Olaniyi EO et al (2019) In-line grading system for mango fruits using GLCM feature extraction and soft-computing techniques. Int J Appl Pattern Recognit 6:58–75
34. Bhargava A, Bansal A (2021) Classification and grading of multiple varieties of apple fruit. Food Anal Methods 14:1359–1368
35. Kumar PS et al (2022) Pomegranate quality analysis and classification using feature extraction and machine learning. In: 2022 6th international conference on electronics, communication and aerospace technology, pp 862–867
36. Jiewen Z (2005) Detection on defects of apples based on support vector machine. J Jiangsu Univ
37. Azarmdel H et al (2020) Evaluation of image processing technique as an expert system in mulberry fruit grading based on ripeness level using artificial neural networks (ANNs) and support vector machine (SVM). Postharvest Biol Technol
38. Patel HB, Patil NJ (2024) Enhanced CNN for fruit disease detection and grading classification using SSDAE-SVM for Postharvest fruits. IEEE Sens J 24:6719–6732
39. Tripathi MK, Maktedar DD (2022) Internal quality assessment of mango fruit: an automated grading system with ensemble classifier. Imag Sci J 70:253–272
40. Naik S, Desai PH (2022) Mango (Mangifera indica L.) Classification using convolutional neural network and linear classifiers. Adv Intell Syst Comput
41. Akter YA, Rahman MO (2017) Development of a computer vision based Eggplant grading system. In: 2017 4th international conference on advances in electrical engineering (ICAEE), 2017, pp 285–290
42. Ayuningtyas D, Suryani E, Wiharto (2021) Identification of tomato maturity based on HIS color space using the K-nearest neighbour method. In: 2021 international conference on artificial intelligence and computer science technology (ICAICST), pp 73–78
43. Trieu NM, Thinh NT (2022) A study of combining KNN and ANN for classifying dragon fruits automatically. J Image Graph
44. Shearer S, Payne F (1990) Color and defect sorting of bell peppers using machine vision. Trans ASAE 33(6):1245–1250
45. Engel BA (1996) Neural networks for quality sorting of agricultural products
46. GHT1193–2021, Tomato. Standards Press of China. Beijing, China, 2021.
47. Muhaemin (2013) Development of an automatic tomato grading machine based on visual evaluation
48. Lu T et al (2021) A generic intelligent tomato classification system for practical applications using DenseNet-201 with transfer learning. Sci Rep 11
49. Opeña HJG, Yusiong JPT (2017) Automated tomato maturity grading using ABC-trained artificial neural networks. Malays J Comput Sci 30:12–26
50. Liu L et al (2019) Design of a tomato classifier based on machine vision. PLoS ONE 14(7):e0219803
51. Deulkar SS, Barve SS (2018) An automated tomato quality grading using clustering based support vector machine. In: 2018 3rd international conference on communication and electronics systems (ICCES), pp 1128–1133
52. Yusiong JPT (2022) A CNN-ELM classification model for automated tomato maturity grading. J ICT Res Appl
53. Li P et al (2023) Tomato maturity detection and counting model based on MHSA-YOLOv8. Sensors (Basel, Switzerland). 23
54. Dao T-M-P, Nguyen N-K, Nguyen V-K (2023) CNN-YOLOv8 - based tomato quality inspection system - a case study in Vietnam. Int J Electr Electron Engin 10(7):31–40

55. Appe SN, Arulselvi G, Balaji GN (2023) CAM-YOLO: tomato detection and classification based on improved YOLOv5 using combining attention mechanism. PeerJ Comput Sci 9
56. Huang Y et al (2024) A fluorescence detection method for postharvest tomato epidermal defects based on improved YOLOv5m. J Sci Food Agricult
57. Amna MW, Guiqiang LI, Muhammad Zuhaib AKRAM MF (2023) Machine vision-based automatic fruit quality detection and grading
58. Huang W, Zhang C, Zhang B (2011) Identifying apple surface defects based on gabor features and SVM using machine vision. In: Conference on control technology and applications
59. Yuan J, Guo Z, Yue D (2011) The apple color grading based on PSO and SVM. In: 2011 2nd international conference on artificial intelligence, management science and electronic commerce (AIMSEC), pp 5198–5201
60. Bhatt AK, Pant D (2015) Automatic apple grading model development based on back propagation neural network and machine vision, and its performance evaluation. AI & Soc 30:45–56
61. Gongal A, Karkee M, Amatya S (2018) Apple fruit size estimation using a 3D machine vision system. Inf Process Agricult
62. Bhargava A, Bansal A (2021) Machine learning-based detection and grading of varieties of apples and mangoes
63. Liang X et al (2022) Real-time grading of defect apples using semantic segmentation combination with a pruned YOLO V4 network. Foods 11
64. Zhao Y et al (2023) Apple grading based on IGWO optimized support vector machine. In: Proceedings of international conference on artificial life and robotics
65. Ji W et al (2023) Apple grading based on multi-dimensional view processing and deep learning. Foods 12
66. Xu B et al (2023) Apple grading method design and implementation for automatic grader based on improved YOLOv5. Agriculture
67. Iosif A et al (2023) Automated quality assessment of apples using convolutional neural networks. INMATEH Agricult Engin

Chapter 5
Recent Advances in Intelligent Harvesting Robots

Yuliang Yun, Xueke An, Dehao Li, and Zhiyong Ma

Abstract Intelligent harvesting robots have become an important trend in contemporary agricultural production, providing efficient and accurate solutions for fruit orchards and fields through the coordinated integration of their "eye-brain-hand-feet" components. This technology encompasses key elements such as information perception, intelligent decision-making, operational execution mechanisms, and autonomous navigation platforms. Information perception involves the use of various sensors to gather environmental and object data, while intelligent decision-making utilizes algorithms and models to autonomously formulate effective decisions. Operational execution mechanisms perform harvesting actions, and autonomous navigation platforms enable robots to navigate autonomously and avoid obstacles in unknown or dynamic environments. However, this technology faces challenges such as environmental complexity, fruit diversity, and autonomous navigation precision. This paper reviews the research progress of this technology over the past five years, providing a comprehensive overview of the current state and future trends. It aims to offer valuable insights to researchers, agricultural producers, and policymakers, facilitating the application and promotion of modern agricultural technology, and advancing agricultural production towards modernization, intelligence, and sustainability.

Keywords Harvesting robot · Environmental perception · Decision control · Autonomous navigation

5.1 Introduction

Modernization and structuring have become a trend in agricultural production methods. Among them, agricultural robots play an increasingly important role in improving agricultural production efficiency and reducing the pressure on agricultural labor. As an important application form, picking robots can not only improve

Y. Yun (✉) · X. An · D. Li · Z. Ma
College of Mechanical and Electrical Engineering, Qingdao Agricultural University, Qingdao 266109, China
e-mail: ylyun1981@163.com

Y. Yun et al. (eds.), *Advanced Sensing and Robotics Technologies in Smart Agriculture*, Smart Agriculture 9, https://doi.org/10.1007/978-981-97-6441-9_5

the production efficiency and quality of agricultural products, but also reduce agricultural labor input, reduce production costs, and promote the sustainable development of the agricultural industry. Autonomous navigation, fruit identification and positioning, operation planning, and picking control are key contents of research on picking robot technology. An efficient, accurate, reliable and low-cost intelligent picking robot with broad application prospects [1].

The intelligent picking robot is mainly composed of four parts: 'eye-brain-hand-feet', which respectively correspond to information perception, intelligent decision-making, operation execution mechanism and autonomous mobile platform. Information perception is the step in which the robot obtains environment and object information. It collects surrounding environment information through various sensors (such as cameras, lidar, thermal imagers), and uses appropriate algorithms and models to process and analyze this information to obtain understanding and awareness of the environment [2–4]. In information perception, the robot can obtain the shape, size, characteristics, distance of the target fruit, and can also identify obstacles, terrain features, etc. Based on the goal and environmental information, the robot uses decision trees, optimization algorithms, or reinforcement learning to make autonomous and effective decisions, coordinating the cooperation between multiple structural units to adapt to different planting situations and achieve overall system optimization [5–8]. The operation actuator is mainly the hardware structure that executes the picking action, including the robotic arm structure and the end-effector. And for different growth conditions of the fruit need to use the appropriate end-effector. The current actuators can be broadly categorized into three types: shear, gripping, and adsorption. Researchers can combine or add modules to meet a wide range of harvesting needs [9]. Autonomous navigation platform is based on localization and autonomous navigation function, which supports robots to move autonomously, make U-turns and avoid obstacles in unknown or dynamic environments [10]. The 'eye-brain-hand-feet' of the intelligent picking robot are coordinated to realize environment sensing, decision-making, execution and movement, which provides an efficient and precise solution for picking work in orchards and farmland.

The purpose of this paper is to discuss the current development status, key technological challenges, and future directions of intelligent picking robotics, and to provide a comprehensive overview of the research progress in the field of intelligent picking robots in the past five years, including sensing and recognition technologies, decision-making and planning algorithms, actuator design and optimization, and mobile platform functions. Meanwhile, this paper will summarize and discuss the challenges faced by intelligent picking robots in practical applications, such as environmental complexity, fruit diversity, and autonomous navigation accuracy, and propose possible solutions and technical strategies to address these challenges. By comprehensively sorting out the current development status and future trends of intelligent picking robotics, this review aims to provide references for relevant researchers, agricultural producers, and policy makers, to promote the application and dissemination of modern agricultural technologies, and to promote the development of agricultural production methods in the direction of modernization, intelligence, and sustainability.

The outline of this paper is shown in Fig. 5.1. In Sect. 5.2, the article details the object and environment sensing techniques, Sect. 5.3 focuses on robotic arm operation planning, Sect. 5.4 describes the fine arm hand structure design, Sect. 5.5 focuses on the implementation of autonomous navigation for mobile platforms, and Sect. 5.6 summarizes the current state of harvesting robotics research and future development prospects.

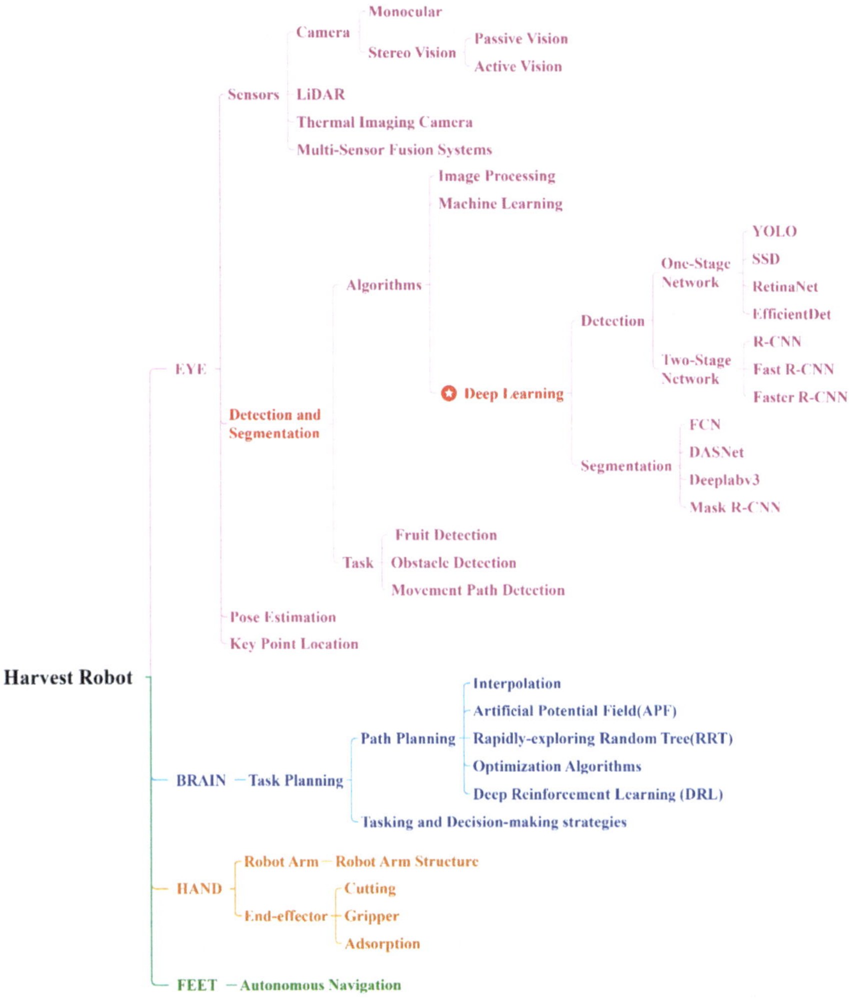

Fig. 5.1 The outline of harvest robot

5.2 'Eye': Object and Environmental Perception Technology

5.2.1 *Perception Sensors*

In the realm of agricultural robotics, sensor technology assumes a pivotal role in furnishing robots with indispensable sensing capabilities and environmental acumen. Sensors facilitate the comprehension of their operational milieu by picking robots, thereby enabling them to execute tasks adeptly through the acquisition of data from the surroundings and its subsequent conversion into actionable information. A diverse array of sensor types, including vision sensors, LIDAR sensors, and thermal imagers, find widespread application in agricultural robots, with the concurrent development of multi-sensor fusion technology. Some representative examples of sensors are shown in Table 5.1.

5.2.1.1 Vision Sensors

In agricultural robotics, visual sensors serve as the eyes of the robot, furnishing it with abundant environmental data. This data encompasses critical attributes such as the shape, size, and color of objects, enabling the robot not only to precisely pinpoint

Table 5.1 Examples of representative sensors

Type		Representative	Detection range (m)	Working principle
Camera	Monocular	Microsoft LifeCam studio Webcam		
	Stereo vision	ZED 2	0.3–20	Neural stereo depth sensing
		Fotonic F80	0.15–5	ToF
		RealSense D435i	0.2–10	Structured light
		Orbbec Astra Pro Plus	0.6–8	Structured light
		Kinect V2	0.4–4.5	ToF
Thermal imager		FLIR ADK™		
LiDAR		Mid-70	Min 0.05	
		UST-10LX	Max 30	
		Velodyne HDL-64E	Max 120	
Multi-sensor fusion systems		RealSense L515	0.25–9	MEMS mirror scanning

and recognize target crops but also to detect and identify obstacles, thereby ensuring safe maneuverability and collision avoidance during operations.

When choosing sensors, it is essential to consider the specific requirements of the task and application scenarios. Monocular cameras excel in acquiring color scene images and furnishing abundant visual data. However, they fall short in providing distance information. For instance, Aljaafreh et al. [11] employed a Microsoft LifeCam Studio Webcam in conjunction with Velodyne LiDAR 3D to obtain distance data. Stereo vision cameras integrate RGB color images with depth data, typically represented in the form of grayscale or distance values. They can be categorized into two types: passive vision and active vision. Passive vision operates on the principle of triangulation, wherein images from two or more cameras are matched and calibrated, and the disparity between pixels is calculated to derive distance information. StereoLabs' ZED series cameras utilize passive vision technology. Vrochidou et al. [4] developed an Autonomous Robot for Grape Harvesting (ARG), equipped with a ZED Mini 3D IMU Camera mounted on the robotic arm. The ZED Mini 3D camera combines stereo vision with inertial measurement unit (IMU) technology to deliver streaming video sequences. Figure 5.2 shows the ZED series cameras, from top to bottom are ZED 1, ZED 2 and ZED Mini camera.

Active vision cameras use Time of Flight (ToF) or structured light technology to obtain distance information. ToF refers to determining the distance between an object and the sensor by measuring the time it takes for light to travel from the emitter to the target object and back to the sensor. Representative cameras include Kinect v2 and Fotonic F series. Structured light refers to projecting a structured light pattern (grating or coding pattern) onto the target object, then using a camera to observe the deformation of these light patterns, and calculating the depth of the target object based on the deformation information. Representative cameras include Intel Realsense series, Microsoft's Kinect v1 and Orbbec Astra. Li et al. [3] utilized an Intel Realsense D435i RGB-D camera as the eyes of the robot to execute picking tasks in the form of eye-in-hand, while an Orbbec Astra 3D camera was embedded

Fig. 5.2 ZED series cameras

(a) Kinect V2 (b) Astra Pro Plus (c) Realsense D435i

Fig. 5.3 Active vision cameras

on the wheeled mobile robot to identify tree trunks for navigation. In the lychee harvesting robot system [12], Kinect V2 is fixedly installed on the platform and used to segment lychee branches. As shown in Fig. 5.3, some representative active vision cameras are presented.

5.2.1.2 LiDAR

LiDAR is a high-precision sensor capable of generating a high-resolution three-dimensional map of the environment, encompassing vital details such as obstacle distance, height, and shape. This detailed map data plays a critical role in facilitating obstacle detection, positioning, navigation, and environmental modeling for the robot. It equips the robot with precise environmental perception capabilities, aiding in optimal path planning and collision avoidance.

Wang et al. [13] installed the 3D LiDAR atop the control cabinet for enabling the robot's autonomous navigation. Mao et al. [14] utilized a positioning unit equipped with the Velodyne VLP-16 line LiDAR, which employs a 16-line LiDAR sensor array to swiftly and accurately capture three-dimensional point cloud data of the environment. The RS-LiDAR-M1 was employed to gather 3D point clouds between rows of orchards [15]. LiDAR emits laser light through rotation or scanning. The greater the number of laser lines, the higher the resolution and accuracy. For instance, the Velodyne VLP-16 LiDAR is equipped with 16 laser lines, enabling it to gather laser beam data simultaneously in 16 directions, thus yielding a more intricate environmental map. However, conversely, the higher the number of lines, the more stringent the requirements for instrument materials, the greater the energy consumption, and the higher the cost. Hence, in practical applications, various factors need to be comprehensively considered to select the appropriate LiDAR.

5.2.1.3 Thermal Imaging Camera (TIC)

A thermal imager is an advanced sensor utilized to detect and record the heat distribution within an environment. Its primary function is to furnish detailed information regarding the temperature of objects. In the context of picking robots, the integration of thermal imaging cameras aids in detecting crucial information such as plant growth, fruit maturity, and environmental temperature fluctuations, thereby offering essential support for the robot's decision-making and action planning.

In certain scenarios, such as low-light conditions and orchard environments under dense tree canopies, Global Navigation Satellite System (GNSS) signals may encounter interference, impacting the robot's positioning and navigation. To address this issue, Jiang et al. [16] combined a thermal imager with a deep learning system, enabling the identification of landmark objects like tree trunks. This integration facilitates effective orchard navigation for the robot across varying lighting conditions. While thermal imagers offer unique environmental thermal information, they often come with a high price tag. Moreover, they may possess lower image resolution compared to RGBD cameras. Nonetheless, thermal imagers are capable of operating in challenging environmental conditions such as low light, haze, or nighttime, thereby retaining significant value and application prospects in specific scenarios.

5.2.1.4 Multi-sensor Fusion Systems

The information gathered by a single sensor is inherently limited. By integrating data from multiple sensors, the shortcomings of individual sensors can be mitigated, the data uncertainty can be minimized, and the comprehension and perception of the environment can be enhanced. The Intel RealSense L515 as shown in Fig. 5.4 is an example of a structured light LiDAR (Light Detection and Ranging) camera developed by Intel. It employs MEMS mirror scanning technology to deliver high-resolution depth images and point cloud data, enabling real-time three-dimensional environment perception and depth measurement. To address the issue of fruit occlusion by leaves and branches of apple trees, Gao et al. [8] mounted an L515 LiDAR camera on the robot chassis in an upward tilted position, thereby increasing the overlap area between the camera's field of view and the robot's working area. Kang et al. [17] proposed a method for semantic segmentation based on the fusion of high-resolution point cloud data from LiDAR and camera. This approach involves scanning the scene with LiDAR to measure the surrounding geometry, which is then combined with color information to generate a comprehensive point cloud. In the case of the ARG (Autonomous Robot for Grape harvesting) developed by Vrochidou et al. [4], four encoders for odometry, an Inertial Measurement Unit (IMU), and a LiDAR sensor are incorporated for positioning. Through the integration of data from diverse sensors, the perception of the environment can be refined, leading to more precise and dependable task execution.

5.2.2 Algorithm Introduction

Traditional image processing techniques, such as threshold segmentation and filtering, are widely used in tasks such as image segmentation. These techniques perform well in simple scenes, but often face challenges in complex environments,

Fig. 5.4 RealSense L515 lidar camera

such as lighting changes and occlusion situations. In contrast, deep learning algorithms have been widely used in the field of target detection and segmentation due to their simplicity and efficiency. Deep learning algorithms use convolutional networks to perform end-to-end learning and feature extraction on images, which can better capture semantic information, thereby improving the accuracy and robustness of detection and segmentation. In addition, machine learning algorithms are also used in target detection and segmentation tasks to a certain extent, such as support vector machines (SVM), K-means, etc. Although their performance in dealing with complex environments is not as good as deep learning algorithms, they are still used in some tasks. plays an important role in specific scenarios. Therefore, deep learning algorithms have become mainstream methods in the field of target detection and segmentation due to their superior performance and flexibility. However, traditional image processing technology and machine learning algorithms still have certain value and application space in specific scenarios and applications. In this section, we will focus on the content of deep learning algorithms.

Deep learning algorithms are mainly divided into one-stage networks and two-stage networks according to stages. The first-stage network includes the YOLO (You Only Look Once) series, SSD (Single Shot Multibox Detector), RetinaNet algorithm, etc. The two-stage network includes RCNN, Fast RCNN, Faster RCNN and further expanded MaskR-CNN, etc. The one-stage network directly implements target detection and positioning through a single neural network model without requiring additional area proposal steps. Its main advantage is that it simplifies the target detection process and improves the detection speed, and is suitable for scenarios with high real-time requirements. Two-stage networks first generate candidate regions (often called region proposals) through one sub-network, and then classify and locate these candidate regions through another sub-network. Compared with one-stage networks, two-stage networks can usually achieve better performance in accuracy, but usually sacrifice a certain detection speed.

5.2.2.1 Yolo

YOLO is a popular real-time target detection algorithm. The earliest version, YOLOv1, was proposed by Redmon et al. [18]. The YOLO algorithm is a one-stage network that treats detection as a regression problem, dividing the input image into grids and predicting the target's bounding box and its corresponding category probability on each grid. Since the inception of the YOLO algorithm, many scholars have applied it in various fields. In robot object detection, the YOLO algorithm is used for fruit detection, obstacle detection, tree trunk detection, and more. It can also be combined with semantic segmentation and point cloud technology to assist with attitude estimation or picking point positioning. Wang et al. [19] proposed YOLOv9 incorporating the concept of PGI (Programmable Gradient Information), which offers comprehensive input information for the target task to compute the objective function. They also introduced a new lightweight network architecture called Generalized Efficient Layer Aggregation Network (GELAN). As shown in Fig. 5.5, YOLOv9 network architecture is presented. This architecture relies on gradient path planning and solely employs conventional convolution to enhance parameter utilization. As of the completion of the author's article, the latest algorithm in the YOLO series, YOLOv10 [20], has been released, introducing consistent dual assignments for NMS-free training of YOLOs and presenting the holistic efficiency-accuracy driven model design strategy.

5.2.2.2 Faster R-CNN

Faster R-CNN (Faster Region-based Convolutional Neural Network) is a classical target detection algorithm proposed by Ren et al. [21]. It is an improved version of the R-CNN (Region-based Convolutional Neural Network) family, which achieves end-to-end target detection by introducing a Region Proposal Network (RPN), while improving detection speed and accuracy. Faster R- CNN is able to handle both target classification and bounding box regression tasks at the same time, and optimizes the network through multi-task learning to improve the performance and generalization ability of the model.

5.2.2.3 SSD

Single Shot Multibox Detector (SSD) algorithm as shown in Fig. 5.6 is a single-stage target detection algorithm proposed by Liu et al. [22]. It draws on the regression idea of YOLO and the anchor mechanism of Faster-RCNN, makes predictions directly on the feature map, and adopts multi-scale feature extraction to improve the detection ability of objects of different sizes, and achieves a good balance between speed and accuracy. The SSD algorithm is fast and accurate, but it still needs to be strengthened in the detection of small targets.

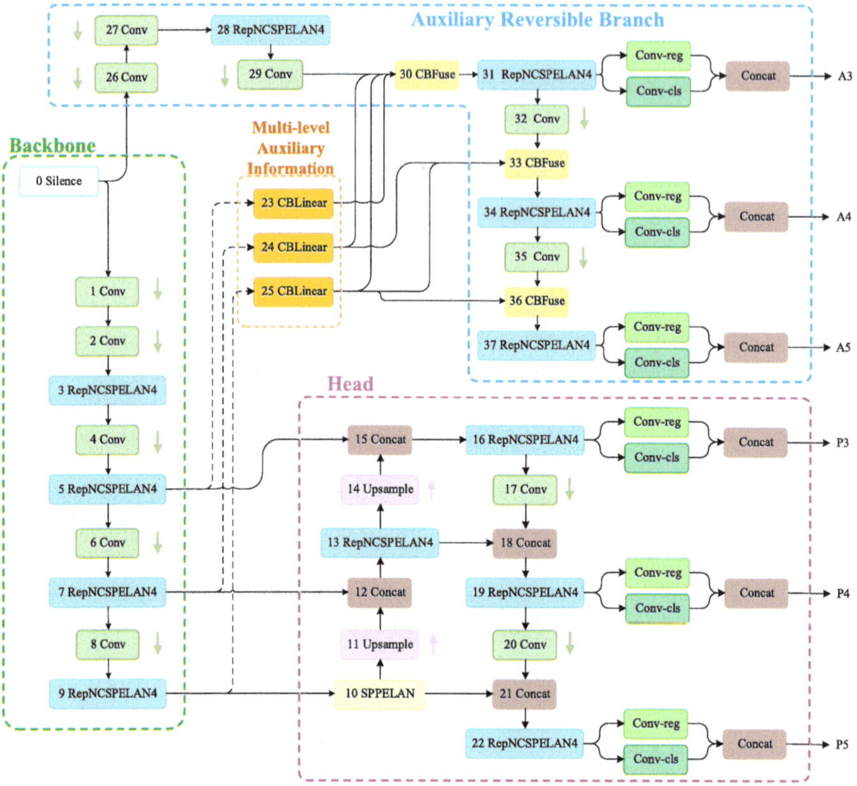

Fig. 5.5 YOLOv9 network architecture

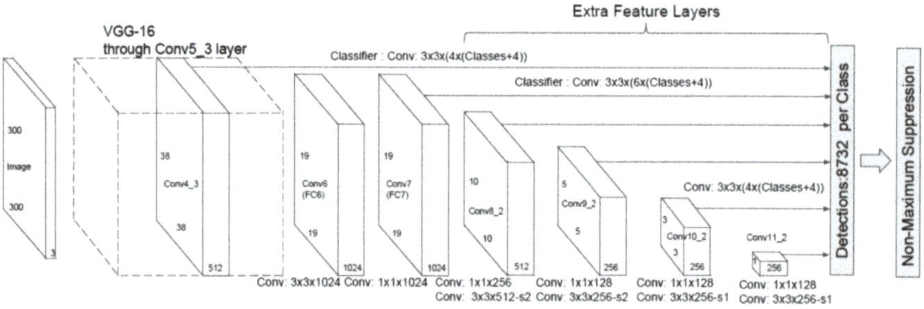

Fig. 5.6 SSD network architecture

5.2.2.4 Mask R-CNN

Mask R-CNN is a deep learning model for instance segmentation task, which is an extended version of Faster R-CNN, proposed by He et al. [23]. Unlike Faster R-CNN, Mask R-CNN adds a segmentation head to each candidate frame for generating an accurate segmentation mask of the target, thus realizing the task of simultaneous target detection and instance segmentation. Mask R-CNN not only accurately detects target objects in an image, but also generates an accurate segmentation mask of each target at the target level Mask R-CNN not only accurately detects target objects in an image, but also generates an accurate segmentation mask for each target at the target level.

5.2.2.5 Deeplabv3

DeepLabv3 is a deep learning model designed for semantic segmentation tasks. It utilizes a deep convolutional neural network architecture and incorporates techniques such as Atrous Convolution and Global Pooling to predict semantic labels for each pixel in an image. As shown in Fig. 5.7, DeepLabv3+ is an enhanced version of this model, used to segment an input image into different semantic regions, assigning each pixel a semantic category.

5.2.3 Object Detection and Segmentation

In the area of target detection, traditional image processing techniques, machine learning, and deep learning models each have their own advantages. Traditional image processing techniques are typically suitable for processing targets with specific shapes or colors and perform better in simple scenes. These techniques include methods such as feature extraction, filtering, and threshold segmentation, which can be used for tasks such as edge detection and shape matching. Wang et al. [24]

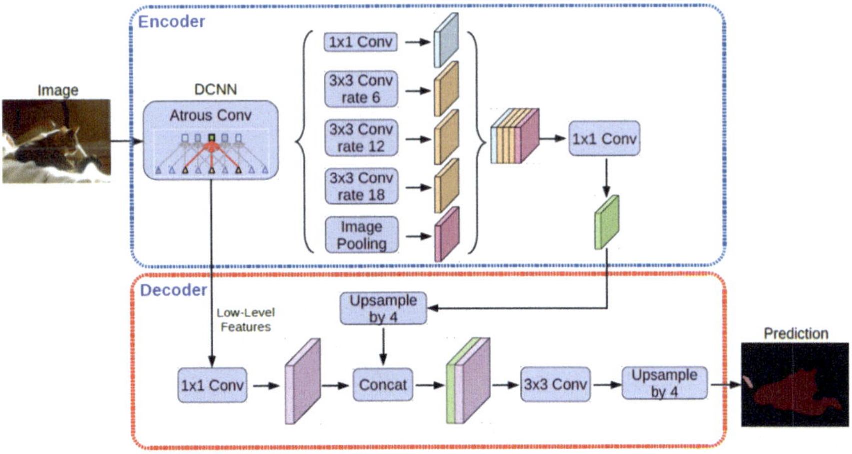

Fig. 5.7 Deeplabv3+ network architecture

applied the OTSU method for apple image segmentation. Zhu et al. [25] used the OTSU algorithm and maximum connectivity domain to obtain target grape images, achieving fast recognition and segmentation of target grapes in 2D space. Ikeda et al. [26] studied the occlusion problem in tomato harvesting using image processing techniques to determine whether the target fruit can be harvested. Dewi et al. [27] used three image processing methods, including image segmentation, edge detection, and blob analysis, to detect and segment fruits. Lou et al. [28] used edge detection and geometric morphology to detect and segment grape bunches. However, image processing-based detection methods are more sensitive to light conditions. Additionally, machine learning methods have been applied in some studies; for example, K-means and KNN methods were used to classify strawberry ripeness in [29, 30], respectively.

In contrast, deep learning models can effectively extract features from images through convolution operations and are currently widely used methods. The implementation of deep learning algorithms requires a large amount of labeled data for training and high computing power for both model training and inference.

5.2.3.1 Fruit Detection

For different picking task requirements, fruit detection algorithms must comprehensively consider various factors, including the use of the scene, lighting conditions, and fruit characteristics. When choosing a suitable algorithm, it is essential to balance accuracy and efficiency, selecting the most appropriate algorithm for the actual needs. Table 5.2 shows some evaluation metrics for object detection.

a. Apple

Apple trees usually grow in orchards or farmland, where they are tall and surrounded by a complex environment with obstacles such as other trees, branches, and leaves.

Table 5.2 Evaluation metrics for object detection

	Matrix	Definition
TP	True Positive	
FP	False Positive	
TN	True Negative	
FN	False Negative	
IoU[a]	Intersection over Union	$IoU = \frac{B_1 \cap B_2}{B_1 \cup B_2}$
P	Precision	$P = \frac{TP}{TP+FP}$
R	Recall	$R = \frac{TP}{TP+FN}$
F1	F1-sore	$F1 = \frac{2 \times P \times R}{P+R}$
AP	Average-Precision	$AP = \int_0^1 P(r)dr$
Accuracy		$\text{Accuracy} = \frac{TP+TN}{TP+FP+TN+FN}$

[a]B1 is the area of the ground truth, B2 is the area of the predicted bounding box

Apple trees produce apples of various shapes, sizes, and colors as they grow, requiring inspection to accommodate different types of fruit. Accurate detection of apples is essential to maximize harvests and ensure effective resource management [31]. Ji et al. [32] proposed and applied ShufflenetV2-YOLOX to detect and localize apples in the orchard environment. The image labels include unbagged apples, bagged apples, and apples at night. The AP (Average Precision) and F1 score of this model are 96.76% and 0.95, respectively, and the semi-precision floating-point (FP16) on Jetson Nano accelerated with TensorRT can reach 26.3 FPS.

Kang et al. [33] proposed an improved deep neural network, DaSNet-v2, for apple detection and semantic segmentation of tree branches. The model employs the lightweight ResNet-18 as the backbone and combines instance and semantic segmentation branches to predict bounding boxes and masks of objects as well as the segmentation of tree branches. The DaSNet-v2 model achieved a recall of 0.85 and a precision of 0.87 for apple detection, 0.866 for fruit segmentation, and 0.757 for dendritic segmentation. In addition, Kang et al. [34] proposed another lightweight network, Mobile-DaSNet, which achieved high performance in detection and instance segmentation with an F1 score of 0.851, a recall of 0.826, a precision of 0.9, and an instance segmentation accuracy of 0.82. Compared to DaSNet-v2, this model has improved accuracy.

Yoshida's team [35, 36] used Mask-RCNN and SSD algorithms for the detection and segmentation of pears (Hosui) and apples (Fuji), respectively. The SSD model's detection results for apples in front lighting achieved a precision of 97.1%, a recall of 98.3%, and an F-measure of 0.977, while the detection results in backlighting were 96.9, 98.0%, and 0.974, respectively. For pears, the SSD model's detection results were slightly inferior, with a precision, recall, and F-measure in front lighting of 95.8, 95.2%, and 0.955, respectively, and in backlighting of 95.5, 97.8%, and 0.968, respectively.

Li et al. [37] proposed a multi-task deep convolutional neural network (DCNN) structure as shown in Fig. 5.8, utilizing both a self-built dataset and the Minnapple dataset. The network employs CSPDarknet as the backbone, and the neck network integrates the feature output of CSPDarknet. The head layer uses the Path Aggregation Network (PAN) for the bottom-up feature pyramid network and includes an additional segmentation branch. In the end-to-end training mode, the network achieved a recall of 74.4%, an average precision (AP) of 67.3%, an accuracy of 92.0%, and an intersection over union (IoU) of 71.0%.

Furthermore, images obtained under low-light conditions often exhibit low contrast and low brightness, negatively impacting fruit identification. Zhang et al. [38] proposed a fast, lightweight, and light-enhanced object detection method called LE-YOLO, which adaptively restores the color information of the input image.

b. Tomato

Tomatoes are typically round in shape, with relatively thick stems. The ripening time of fruits on the same tomato cluster varies slightly, and their skin appears shiny. In unstructured tomato cultivation environments, leaf occlusion and changes in lighting conditions can affect detection. Deep learning-based target detection offers

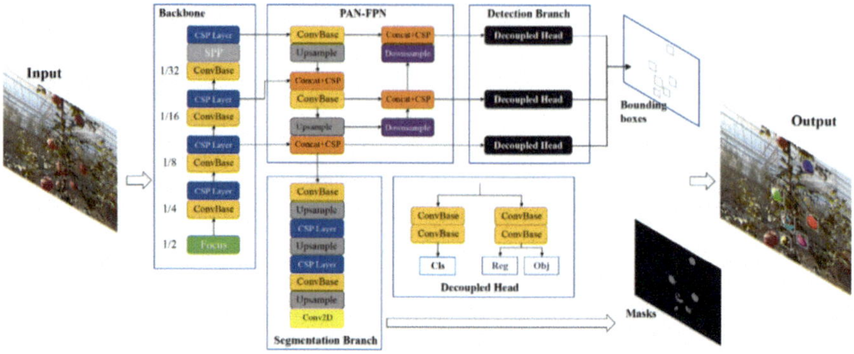

Fig. 5.8 The multi-task DCNN of fruit identification [37]

advantages such as short detection times and high accuracy, meeting the demands of real-time detection in complex environments [39, 40]. Mbouembe et al. [41] improved the YOLOv4-tiny model, achieving an accuracy of 96.3%, a recall rate of 95%, an F1 score of 95.6%, an average IoU accuracy of 82.8% for IoU of 0.5–0.95, and a detection time of 1.9 ms/pic. Zheng et al. [42] developed RC-YOLOv4, a tomato detection model suitable for complex natural environments, with an average detection accuracy of 94.44%. Tsai et al. [43] proposed three object classification models based on Yolov5m combined with BoTNet, ShuffleNet, and GhostNet convolutional neural networks (CNNs), achieving detection accuracies of 94, 95, and 96%, respectively.

In terms of model lightweighting, Yang et al. [44] proposed a lightweight YOLOv8s model for automatic tomato detection. Compared to the original version, the model size was reduced by 4 M, with lower loss and achieving a 93.4% mean average precision (mAP) on their self-built dataset. Wang et al. [45] introduced the Small MobileNet YOLOv5 (SM-YOLOv5) detection algorithm and model, which has a compact size of only 6.33 MB, an accuracy rate of 97.8%, and a recall rate of 96.7%. Wang et al. [46] presented a lightweight real-time detection algorithm, I-YOLOv5n, applied to cherry tomatoes, with a model size of 4.4 MB and an mAP of 95.2%.

c. Strawberries

Strawberries are typically heart-shaped fruits, often red or nearly red, although certain varieties may exhibit white or pale pink hues. They possess tiny granules on their surface, which become glossy when ripe. Strawberry plants feature short stems and typically droop in plantations. The leaves are slightly larger in comparison to the fruit and often cast shade over the fruit or stem. Due to the strawberries' small size, Tao et al. [47] introduced a Biformed attention mechanism in the YOLOv5s model, reducing computational effort and applying it to recognize strawberry ripeness.

To address the challenge of inadequate detection of small strawberries and the occurrence of false positives due to the complex background of strawberry images, Luo et al. [48] introduced an enhanced strawberry recognition approach based on the YOLOv8n model. Modifications were made to the small object detection layer in this model, alongside the integration of a global attention mechanism in the neck. The effectiveness of the model was verified using a self-constructed strawberry dataset. Cao et al. [49] proposed a straightforward framework called StrawSeg for directly segmenting each strawberry in a single-shot manner, without relying on object detection. Model evaluation on the open-source dataset StrawDI_Db1 demonstrated that the model can achieve a commendable balance between accuracy and speed on low-power devices.

d. Zanthoxylum

Zanthoxylum fruits grow densely and cluster on the branches of the tree. Zanthoxylum trees have intertwined branches and dense foliage, which may obscure the fruit or make observation challenging. To address the complexities of this growth environment, Xu et al. [50] proposed a lightweight YOLOv5s model tailored for detecting Zanthoxylum, which exhibits multi-clustering characteristics. This model adopted the FReLU activation function, introduced an improved CBF module based on the CBH backbone module, and utilized a CBF-based Specter module to replace the bottleneck CSP module. With a model size of 14.0 M, it was deployed on NVIDIA Jetson TX2, achieving an average detection speed of 0.072 s/pic and an mAP value of 94.5%. Similarly, Guo et al. [51] presented an enhanced YOLOv5 model, named CA-DCNv2-YOLOv5, specifically designed for peppercorn detection. This model integrates the coordinate attention module and the deformable convolution module into the YOLOv5 network, resulting in a model size of 14.7 M. It achieves an average detection time of 11.2 ms and a detection speed of 89.3 FPS.

e. Asparagus

Liu et al. [52] proposed a DA-Mask RCNN model that uses the depth information in the region proposal network to perform detection of green asparagus. This model is a combination of a deep residual network and a feature pyramid network as a backbone. Five datasets were established in the study. Set: sunny weather (S), sunny but overshadowed conditions (O), cloudy weather (C), daytime greenhouse conditions (D) and nighttime greenhouse conditions (N). Table 5.3 shows the performance of DA-Mask RCNN model under different weather and light conditions.

Hong et al. [53] proposed an improved YOLOv5 algorithm to identify and detect asparagus in complex environments. The training, validation, and test dataset are

Table 5.3 Performance of DA-Mask RCNN model under different weather and light conditions [52]

Weather and light conditions	Precision	Recall	F1-score
Sunny	0.941	0.930	0.935
Sunny but overshadowed	0.887	0.863	0.875
Cloudy	0.952	0.950	0.951
Greenhouse-daytime	0.993	0.971	0.982
Greenhouse-nighttime	0.959	0.934	0.946
Average	0.946	0.930	0.938

divided in a ratio of 8:1:1. The precision and recall of YOLOv5-CB are 97.88 and 96.79% respectively, and the detection speed is 31 FPS. Compared with YOLOv5, the precision and accuracy have been improved, and the speed has slightly decreased.

f. Grapes

Grapes are known for their fragility, typically growing at higher positions on vines and clustering together. Different grape varieties exhibit distinct characteristics, leading to variations in detection methods. Wei et al. [54] proposed a grape detection method based on YOLOv3, while Guo et al. [55–57] introduced improvements to the YOLOv4 algorithm, with models validated on the WGISD dataset. Wang et al. [58] integrated dual-channel feature extraction attention (DCFE) into YOLOv5s, incorporated dynamic snake convolution (DS-Conv) in the backbone, and developed GrapeDetectNet (GDN), demonstrating robustness in their self-made dataset, albeit limited to two grape varieties: Shine Muscat and Kyoho. Coll-Ribes et al. [59] utilized Mask R-CNN on RGB-D four-channel images to detect and segment various grape stems and bunches. Notably, the depth channel in these images was obtained via monocular depth estimation, which, according to test results, enhances foreground–background contrast, thereby facilitating reliable detection outcomes.

g. Others

Aljaafreh et al. [11] curated a dataset comprising 1.2 k images of olive fruits on trees and evaluated various versions of YOLOv5 and YOLOR. YOLOv5s demonstrated superior performance in real-time detection of olive fruits on trees. YOLOv7 was employed for the detection of Camellia oleifera fruit [60], involving fruit detection and extraction of fruit contours to determine their center of mass. Wang et al. [19] utilized RGB-D sensor data to generate an octree representation of plants, annotating regions of interest (ROIs) for fruit identification. López-Barrios et al. [61] applied Mask R-CNN to detect both fruits and pedicels of bell peppers, focusing primarily on sweet peppers with colors similar to the background. The overall accuracy rate was 78.16%, with a recall rate of 66.86% and an F1 score of 71.89%. Su et al. [62] proposed a lightweight cucumber recognition model named YOLOv5s-Super, designed to mitigate interference from near-color backgrounds in greenhouse environments.

5.2.3.2 Obstacle Detection

In contemporary orchard or farmland environments, picking robots often encounter challenges such as obscured fruit due to branches, foliage, and other obstacles. These hindrances can result in decreased harvesting efficiency and inaccurate fruit recognition, while also raising the risk of damaging the picking machinery. Therefore, addressing these challenges and minimizing the impact of obstacles on the picking process necessitates obstacle detection. The objective of obstacle detection is to identify and precisely localize various obstacles within an orchard or farmland, distinguishing them from the target fruit. Effectively identifying and localizing obstacles enables the picking robot to navigate around them, mitigating the risk of collisions, minimizing robot damage, and enhancing both picking efficiency and accuracy.

Li et al. [63] employed an instance segmentation network to extract geometric information and localize the target. Their dataset comprises two components: a self-constructed dataset and the Minnapple dataset. The self-constructed dataset includes images captured under various lighting conditions, occlusion scenarios, and time periods. They conducted experimental comparisons of eight classical segmentation networks and ultimately selected the YOLACT++ network with ResNet101 as the backbone. Following network segmentation, a point cloud was generated and clustered using DBSCAN (Density-based spatial clustering of applications with noise), through which the points corresponding to the target fruits were aggregated. Moreover, they estimated the radius and center position of the fruits to derive the robot's proximity direction vector, computed from the fruit center towards the center of mass of the resulting clusters. Wang et al. [64] utilized a combination of image processing techniques and the YOLOv8 algorithm to detect obstacles surrounding the picking point, achieving 100% accuracy in obstacle recognition. Binocular vision was employed for picking point localization, and active obstacle removal was developed. The YOLOv8-seg model exhibited a combined performance of 88.1% accuracy, 86.0% recall, 87.04 F1 score, and 78.1 mAP.

Duet al. [65] introduced an end-to-end multi-task convolutional neural network called YOLO-MCNN (multitask convolutional neural Networks) depicted in Fig. 5.9,

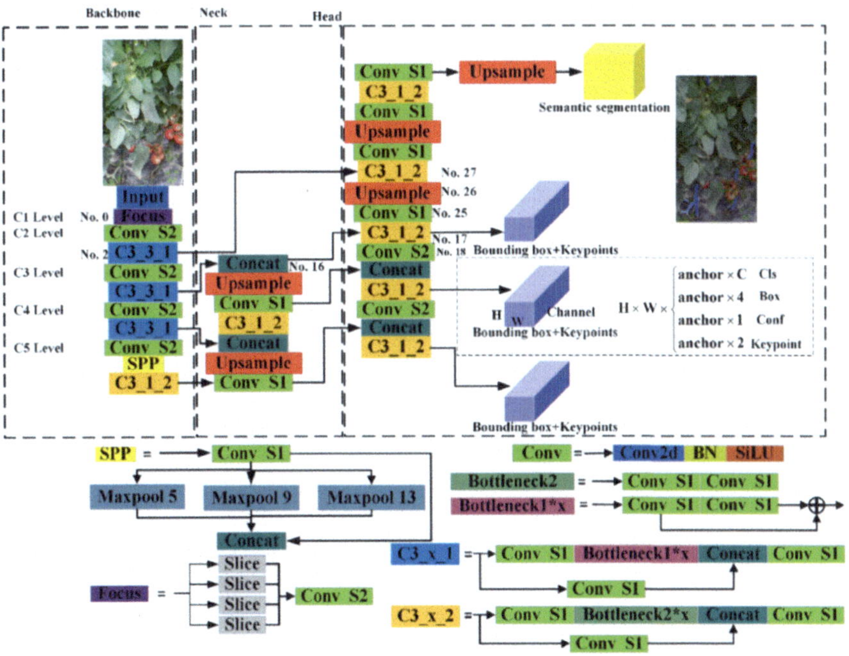

Fig. 5.9 YOLO-MCNN network structure [65]

capable of simultaneously performing target positioning, attitude detection, and obstacle semantic segmentation. The study proposed four feature fusion strategies and semantic segmentation branch configurations. By comparing their performance, the optimal model was determined among the four methods. In this configuration, the semantic segmentation branch was integrated into the 17th layer, with additional C3_1_2 and C3_1_2 Conv modules. Three Upsample layers were utilized to restore the feature map to the original image size for semantic segmentation. A 1×1 convolution was applied in the last semantic segmentation layer for feature point classification, and feature fusion was achieved by combining layer 2 with layer 27. The YOLO-MCNN model was deployed in tomato picking, with the target of semantic segmentation being the tomato stem—a small and challenging target to segment. Consequently, it was imperative to investigate whether the accuracy and effectiveness were impacted when detection and segmentation tasks shared a backbone network.

5.2.3.3 Movement Path Detection

In the autonomous navigation of robots, automatic traversal between fruit trees based on machine vision is deemed a dependable approach. This method involves detecting rows of fruit trees or their trunks to establish the robot's path. Path detection typically employs devices like visual sensors, lidar, or drones to gather information about the vehicle's surrounding environment, enabling real-time identification of potential obstacles, hazardous zones, or other safety risks. Through analysis and assessment of the collected data, the prevailing driving conditions and efficiency of the current path can be promptly ascertained and optimized to avert collisions and accidents, thereby ensuring vehicle driving safety. The general flow of navigation line extraction is shown in Fig. 5.10.

Kim et al. [66] proposed a patch-based automatic path detection method for orchards. This approach utilizes a patch-based convolutional neural network for image classification to accomplish path segmentation, coupled with boundary line

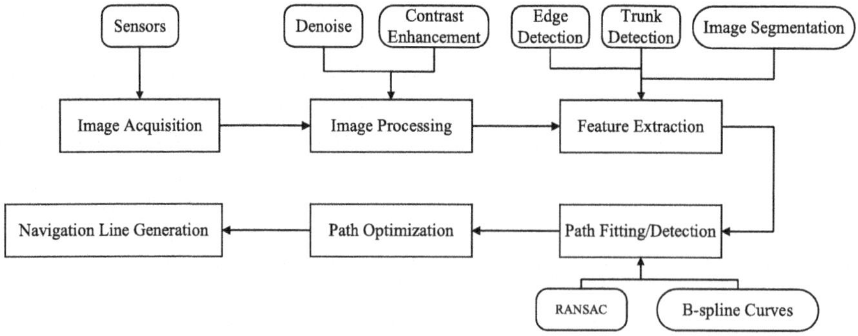

Fig. 5.10 General process of navigation line extraction

determination technology for path detection. Experimental evaluation revealed that employing a patch size of 96 × 96 yielded optimal performance, achieving a classification accuracy of 0.93, an IoU of 0.75, and a processing time of 0.111 s. While this method demonstrates efficacy in semi-structured orchards, its performance in unstructured environments remains to be validated.

U-Net is a deep learning network architecture commonly employed for image segmentation tasks. Initially developed for biomedical image segmentation, it has found widespread application in various domains, including path detection within the realm of autonomous navigation. Han et al. [67] introduced a road semantic segmentation approach based on U-Net. This method involves binarizing the segmentation outcomes, extracting road edge information, generating fitted midpoints, and ultimately producing navigation paths through multi-segment three times B-spline curve fitting to the fitted midpoints. In experiments, the proposed method exhibited reduced susceptibility to lighting conditions, with an average pixel error of 9.5 pixels and an average distance error of 0.044 m.

Mask R-CNN also has a place in path detection. For instance, in addressing issues like incomplete positional information for monocular visual navigation and low localization accuracy of fruit trees, Bi et al. [68] employed an enhanced Mask R-CNN method for segmenting roads and trunks. They applied a Hough transform to the convex hull of the road mask to extract road boundaries and vanishing points, constructing a geometric model for positional-road imaging. Experimental results indicate that the improved Mask R-CNN model achieves an average accuracy of 0.564 for border regression, 0.559 for segmentation, and an average inference time of 110 ms. To address challenges in autonomous travel and turning between rows for visually navigated robots, Yang et al. [69] proposed a Mask R-CNN-based method for extracting navigation lines and an enhanced Random Sampling Consistency (RANSAC) algorithm for tree line extraction. Experimental results demonstrate that the model achieves an average segmentation accuracy and bounding box detection accuracy of 97.0% across six orchard scenarios under various lighting, weed, and weather conditions, with an average deviation of navigation target point extraction not exceeding 5.3%, a tree line detection accuracy of no less than 87%, and an average deviation of the vehicle from the center of the road after turning of 7.8 cm.

Badeka et al. [70] proposed EfficientDet-D0 for real-time detection of grapevine trunks, laying the foundation for navigation of self-service harvesting robots. Zhou et al. [71] employed the YOLOv3 model to detect tree trunks and trees, achieving a mean Average Precision (mAP) of 92.11%. Additionally, they utilized the least squares method to fit the baseline of fruit trees on both sides to determine the navigation route.

5.2.4 Pose Estimation

Fruit pose estimation is utilized to precisely determine a fruit's position and orientation in 3D space through the analysis of its location, orientation, and pose information. In orchard settings, the shape and position of fruits can be influenced by factors such as trees, foliage, and other environmental elements, necessitating consideration of both environmental data and the intrinsic morphological characteristics of the fruits. Recent research has highlighted the significant benefits of employing RGB-D cameras for fruit pose estimation. These cameras not only capture high-quality color images but also gather depth information for each pixel, facilitating more accurate estimations of fruit position and morphology. Additionally, some studies have leveraged LiDAR sensors to acquire distance data for fruits. LiDAR offers precise distance measurements and is unaffected by variations in light, making it a valuable tool for attitude estimation in specific environments.

FCN, Fully Convolutional Network, proposed by Shelhamer et al. [72], is a deep learning model designed for image segmentation tasks. It classifies image pixels into different categories, enabling precise segmentation at the pixel level. FCN comprises convolutional layers exclusively, facilitating its ability to process input images of varying sizes through multi-scale information fusion and end-to-end training. Lin et al. [73] introduced a method for detecting multiple three-dimensional (3D) line segments to reconstruct tree branches and estimate fruit 3D poses using fruit center positions and branching details. In their study on guava-harvesting robots, equipped with an RGB-D camera Kinect V2, fruits and branches were segmented using the FCN model, and Euclidean clustering was applied to extract individual fruits from the fruit point cloud. The precision and recall of guava fruit detection were 0.983 and 0.948, respectively, with a 3D position error of $23.43^\circ \pm 14.18^\circ$ and an execution time of 0.565 s per fruit. However, cases where branches are unrecognized or guava is misclassified as leaves pose challenges for accurate pose estimation.

PointNet, introduced by Qi et al. [74], is a deep learning model specifically designed for processing disordered point cloud data. PointNet structure is shown in Fig. 5.11. As the first end-to-end deep learning architecture capable of directly handling point cloud data without preprocessing or conversion into alternative network forms, PointNet exhibits rotational and substitution invariance. This property allows it to effectively process unordered point cloud data irrespective of the arrangement order of the points. PointNet works by designing two modules, global feature learning and local feature learning, to PointNet is designed to learn both global and local features of the point cloud by designing two modules: global feature learning and local feature learning. The global feature learning module is used to capture the overall structure and global features of the point cloud, while the local feature learning module is used to capture the local features of each point in the point cloud. Kang et al. [34] utilized PointNet for processing point clouds and estimating grasping, leading to the development of a robotic harvesting system with a success rate of 0.8 and a cycle time of 6.5 s. Kang et al. [75] a geometric perceptual network named A3N (Apple 3D Network) was proposed to identify regions of interest

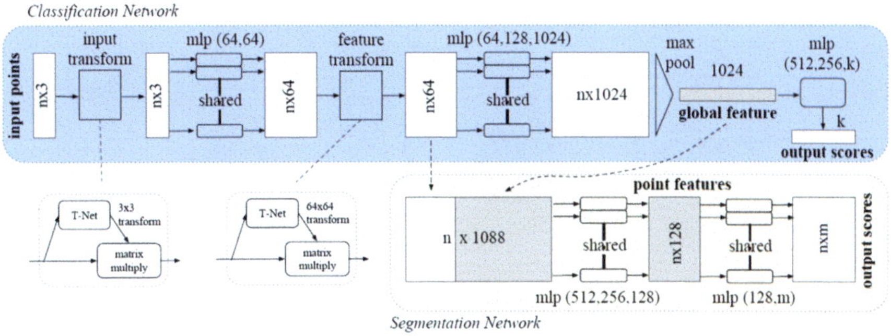

Fig. 5.11 PointNet architecture

(ROIs) from RGB images. Furthermore, the PointNet model was enhanced to perform bounding box regression on the point cloud of each apple. The resulting A3N-640-R101PANet achieved an accuracy of 0.893 on IoU mask. Kang et al. [17] presented a 3D segmentation approach based on PointNet, referred to as Fuse-PointNet++. This method achieved a segmentation accuracy of 0.88. Data acquisition was conducted using a LIDAR-Camera setup.

Kang et al. [76] applied Dasnet to detect fruits and segment branches, while also calculating the grasping posture of each fruit using the Hough transform. Baisa et al. [77] utilized circular Hough transform for mushroom detection and proposed a 3D pose estimation algorithm incorporating both global registration and local refinement registration methods. The global registration method aligns point cloud datasets acquired from different perspectives or times into a rough global reference frame, while the local refinement registration method further optimizes the registration results in the local area to achieve more precise alignment of point cloud data. Retsinas et al. [78] employed information from multiple stereo cameras to estimate the 3D pose of mushrooms from multi-view point clouds, albeit at a higher cost.

Yin et al. [79] leveraged the Mask R-CNN network to segment grapes from RGB images, extracting grape cluster point cloud data. Subsequently, they utilized the RANSAC algorithm to fit the grape clusters into cylindrical models to achieve grape attitude estimation. Wagner et al. [80] proposed a network framework capable of estimating the rotational configuration of rotationally symmetric objects, with experiments conducted on the strawberry public dataset. Kim et al. [81] introduced a deep learning network pipeline called deep-tomatos, which can simultaneously estimate six-dimensional (6D) poses (3D transformation and 3D rotation) with an average accuracy of up to 96%. However, this 6D pose estimation process requires a significant inference time and needs to address challenges in implementing pose updates. Park et al. [82] proposed a method based on computer vision detection to rapidly estimate the posture of cucumber pedicles, as illustrated in Fig. 5.12.

Le Louëdec et al. [83] proposed a method for predicting the three-dimensional orientation of fruit from a single-view image. The orientation of the strawberry was estimated based on the detection results of two key points: one at the top and

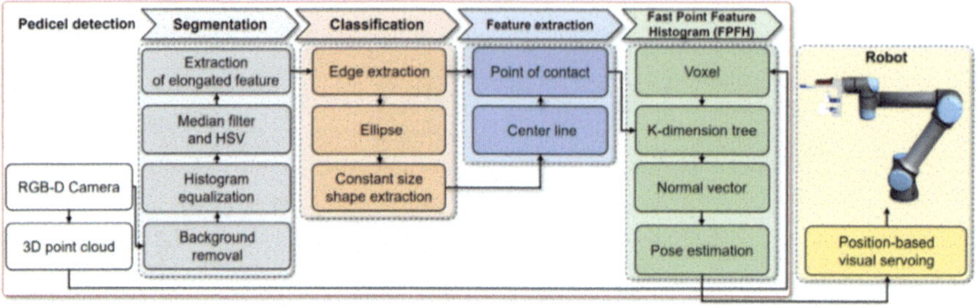

Fig. 5.12 Process flow of cucumber pedicel detection [82]

the other at the tip of the strawberry, as well as the overall shape of the strawberry. However, this method does not account for occlusions of key points. Zhang et al. [84] combined a prior geometric model, a cascaded multi-task network, and three-dimensional reconstruction to propose a method for three-dimensional attitude detection of tomato bunches. This provides a data basis for target positioning and path planning of picking robots, enabling lossless harvesting.

5.2.5 *Keypoint Detection and Localization*

Keypoint detection and localization involve identifying the precise position of fruit on the plant by analyzing the plant's growth pattern, the fruit's shape, and the relative positions of plant parts. This provides accurate target information for picking robots to locate and execute the picking action accurately. Advances in deep learning and convolutional neural networks have led to significant breakthroughs in image processing and computer vision, offering more reliable methods and tools to improve the accuracy and efficiency of picking robots.

In the study of key point localization for tomatoes, Yoshida et al. [85] utilized SVM to perform voxel segmentation based on information obtained from an RGB-D camera. Rong et al. [86] proposed a semantic segmentation model based on Swin Transformer V2, achieving accurate identification of tomato fruits, sepals, and stems. They used image refinement algorithms, deblurring algorithms, and sepal position constraints to identify picking points on the tomato side stems, achieving an accuracy and recall rate of 97.43 and 86.04%, respectively, for the picking point identification algorithm. Kim et al. [87] successfully employed an end-to-end method to detect key points in images taken in a tomato greenhouse, but their work was limited to predicting the 2D pose of the pedicel connecting a single tomato to the truss. In clustered tomatoes, mature tomato clusters are arranged in a straight fishbone pattern on a regular plane, with similar orientation, size, and even distribution. Bai et al. [88] combined machine learning, Hough circle detection, and geometric analysis to propose a visual algorithm for target recognition, pedicel estimation, and picking point localization based on the shape and growth characteristics of clustered tomatoes. However, the detection performance of this algorithm is affected by factors such as time, space, and imaging noise. Ci et al. [89] applied the Keypoints R-CNN network as a key point detector for tomatoes.

The distribution of lychee fruit on trees is irregular, often forming clusters. Due to the high freshness requirements of lychees, the fruit and the connected stems need to be harvested together. Based on this, Peng et al. [90] utilized Deeplabv3+ to segment lychee branches. Li et al. [12] employed Deeplabv3 for semantic segmentation to extract the main stem of the lychee clusters. A non-parametric density-based clustering method was then applied to cluster the 3D spatial pixels of the branch skeleton image, determining the clusters of the same lychee. Principal component analysis was used to fit individual clusters and locate the corresponding fruit branches. Qi et al. [91] combined the YOLOv5 and PSPNet models to segment the main stem of the lychee, and the pixel coordinates of the picking points were obtained through coordinate transformation.

Furthermore, Wang et al. [64] utilized the YOLOv8-seg model to segment lychee and branches, determine their relationships, and perform binarization. The branch was divided into ten segments along its height, and the width of each segment was compared to determine the picking point. The coordinates of the picking point were then obtained from the disparity map generated by a stereo camera. Wang et al. [92] proposed a distal–proximal coordinated control strategy for clustered fruits. YOLOv5 was utilized for identifying individual litchis from a distance, while Mask R-CNN was employed for close-range segmentation of trunks and branches. Based on the segmentation results, the picking points were then determined (Fig. 5.13).

Similarly to litchis, grapes also grow in clusters, but with shorter stems. Jin et al. [93] proposed a method for accurately locating grape harvesting points by combining

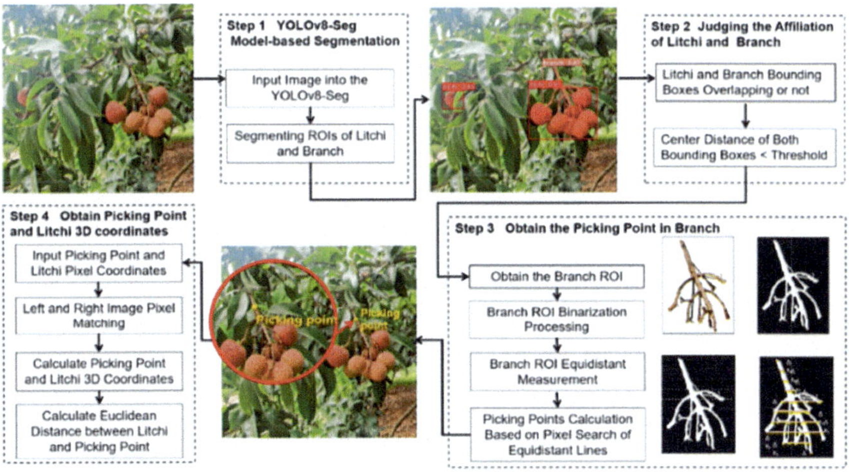

Fig. 5.13 Algorithm flow of locating picking points [64]

distant and near-depth data features from trellis-cultivated grape vines. The robot detects and roughly locates grape clusters from a distance, guiding the end-effector close to the selected grape cluster for further detection and key point determination. Experimental results indicated that incomplete grape cluster images and lower bottom points in the near field can lead to misjudgments of the cluster's bottom points. Moreover, strong light can result in incomplete stem point cloud capture by the camera. Large position deviations of the end-effector may also cause collisions or task failures. Jin et al. [94] proposed another distant and near stereo vision system for detecting grape bunches and stems. For grapes grown on trellises, Xu et al. [95] introduced a rapid multi-grape picking point localization method based on synchronous inference, achieving an average Euclidean distance localization error of 7.69 mm between actual and predicted picking points. Zhao et al. [96] proposed a lightweight model, YOLO-GP, for simultaneous detection of grape clusters and picking points, achieving an mAP of 93.27% on a custom grape image dataset, with a picking point detection distance error of less than 40 pixels. Although it reduced background interference, it did not consider the impact of different end-effectors on the estimation and calculation of picking points, potentially causing inaccuracies in grasping the grape stem. Zhu et al. [97] introduced Yolov5-CBAM-Fourth Detection Layer-Decoupled (Yolov5CFD), which enhanced small object detection and used geometric methods to quickly and accurately locate picking points for table grapes. Zhu et al. [25] designed a grape picking point localization device based on infrared sensors, achieving precise three-dimensional spatial localization of picking points for table grapes.

China is the world's largest tea producer and a significant consumer of tea. Unlike fruit harvesting, tea leaf picking involves the discernment between new buds and older leaves at the top of the tea plant, as the younger buds are of higher quality and value. Lu et al. [98] proposed a method for segmenting fresh tea buds and localizing picking points based on the high-resolution network (HRNet_W18). Yan et al. [99] employed an enhanced DeeplabV3+ semantic segmentation algorithm to segment the foreground of tea buds, extracting the largest connected domain after binarization. Subsequently, the Shi-Tomasi algorithm was applied to calculate the corners of the largest connected domain, with the lowest corner marked as the picking point. Yan et al. [100] introduced a harvesting point localization model, MR3P-TS, based on Mask R-CNN, which determines the picking point according to the direction of the shoot axis and recommends the optimal cutting angle. Meng et al. [101] proposed an improved PSP-net to detect and segment areas for picking tea buds, using the centroid of the detected box as the picking point. Zhang et al. [102] combined DeepLabv3+ and YOLOv7 to propose the MDY7-3PTB model, which segments, detects, and then locates tea buds. Due to the variability and irregularity in tea bud growth patterns, depth information is essential for the three-dimensional localization of tea bud picking points. Zhu et al. [103] introduced the YOLOv5s-Tea model, which generates a tea bud point cloud based on detection results and utilizes principal component analysis to obtain the 3D coordinates of the picking points.

Zheng et al. [104] proposed a mango image segmentation and key point detection method that effectively segments and detects key points in small to medium occluded or overlapping mango images, as well as medium to large mango images. However, this method does not address the issue of locating picking points in mango clusters. Zhang et al. [105] enhanced the YOLOv5s network by integrating stem segmentation to locate picking points in entire mango clusters. Hu et al. [106] utilized a ZED stereo camera, triangulation, and a neural network for the three-dimensional localization of strawberries, although the study did not specifically focus on picking point localization. Song et al. [107] employed DeepLabV3+ based on Resnet-101 to segment calyx, branches, and threads in kiwi canopy images, thereby enhancing the picking success rate. Zeng et al. [108] applied an improved Mask R-CNN to segment node regions of mulberry leaves and utilized a refinement algorithm to skeletonize the node regions for picking point localization. Zhou et al. [109] utilized YOLOv7 detection and PSP-Ellipse segmentation to identify and locate dragon fruit, detecting both the head and root endpoints. The distance error for endpoint localization based on ellipse fitting was 39.8 pixels, and the angle error was 4.3 degrees.

5.3 'Brain': Intelligent Decision-Making

5.3.1 Path Planning

Robot arm path planning involves determining the trajectory in space that a robot arm will follow to accomplish a designated task. This intricate process considers various factors, including the dynamic capabilities of the robotic arm, limitations within the workspace, strategies for avoiding obstacles, optimization objectives, and other pertinent constraints.

a. **Interpolation**

   Interpolation is a critical technique in robot arm path planning, encompassing methods such as linear interpolation, circular interpolation, spline interpolation, and others. Linear interpolation is typically employed when the robot arm moves along a straight path with stable speed and acceleration. Arc interpolation, on the other hand, is utilized for scenarios where the robot arm executes arc motions, such as moving along an arc path between two target points. This method is particularly effective for tasks involving circular workpieces. To achieve smoother paths, spline curves are frequently employed to connect target points. Common spline interpolation methods include cubic spline interpolation and Bezier curve interpolation.

   Cao et al. [110] utilized B-splines for trajectory planning, aiming to achieve smoother motion paths for robotic arms. Lan et al. [111] established a time-energy-disturbance optimization mathematical model, leveraging a high-order B-spline interpolation technique to enhance trajectory smoothness. To prevent the manipulator from approaching the fruit in an incorrect direction, Wang et al. [13] introduced a method involving the addition of interpolation points to constrain the manipulator. This ensures that the planned path of the manipulator aligns with the direction of the gripper opening, thus mitigating the risk of collision between the gripper and the fruit. However, in complex workspaces where obstacles may obstruct the movement of the robotic arm, it becomes imperative to consider the integration of other path planning methods.

b. **Artificial Potential Field (APF)**

The Artificial Potential Field (APF) algorithm is a classical approach to robot path planning, primarily used for obstacle avoidance and navigation tasks. In this algorithm, the robot is analogized as a charged particle navigating through an artificial potential field, where obstacles in the environment are represented as charged objects. The direction of the robot's movement is determined by calculating the forces acting on it within this potential field. The APF algorithm relies on two main forces: repulsive and gravitational. The repulsive force pushes the robot away from obstacles to prevent collisions, while the gravitational force attracts the robot towards the target destination. While this algorithm is appreciated for its simplicity and responsiveness to environmental changes, it also has drawbacks, such as susceptibility to locally optimal solutions and limited adaptability to dynamic obstacles. Zhang et al. [112] proposed an improved version of the APF algorithm based on the speed field concept. This enhancement, termed the T-APF algorithm, introduces the notion of gravitational and repulsive velocities. Additionally, the authors incorporate the Jacobi matrix and its inverse matrix to establish a differential velocity function for joint motion, aiming to enhance collision avoidance. Despite these advancements, the proposed method still exhibits limitations, particularly in parameter optimization, necessitating further research in this area.

c. **Rapidly-Exploring Random Tree (RRT)**

The Rapidly-exploring Random Tree (RRT) is a path planning algorithm that relies on random sampling and rapid exploration to efficiently navigate through high-dimensional spaces, offering strong convergence performance. The RRT algorithm achieves this by systematically exploring the entire state space through the random generation of nodes and the subsequent connection of these nodes to form a tree structure. Throughout this exploration process, RRT prioritizes obstacle avoidance to ensure that the generated paths remain free from collisions with obstacles. As a result, RRT algorithms find extensive application in robotic arm path planning, particularly in addressing obstacle avoidance challenges within complex working environments. Ma et al. [5] utilized the RRT algorithm for robotic arm path planning by integrating target fruit attitude information and continuously updating fruit information upon completing each picking task until all tasks are accomplished. Tian et al. [113] proposed an RRT algorithm that incorporates target gravity and further optimizes generated paths using genetic algorithms and flattening algorithms. Ye et al. [114] employed an improved Adaptive Weighted Particle Swarm Optimization (APSO) algorithm to obtain collision-free picking poses. Additionally, they utilized an optimization algorithm based on the Bi-RRT algorithm (AtBi-RRT) to rapidly derive suitable collision-free picking paths. In laboratory scenarios, the success rate of path determination reached 100% during lychee picking.

d. **Optimization Algorithms**
Commonly employed optimization algorithms in path planning encompass the ant colony algorithm, genetic algorithm, simulated annealing algorithm, and particle swarm algorithm. Zhang et al. [7] delved into automatic path planning for safflower picking and introduced an enhanced version of the ant colony algorithm (ACA). Simulation outcomes demonstrated that the path derived through the enhanced ACA reduced the number of picking points by three, thereby shortening the total length by 74.32% and reducing picking time by 0.957 s. Zhang et al. [115] proposed an optimal sequential ant colony optimization algorithm (OSACO), transitioning the picking path planning from a point-to-point approach to continuous picking. Gao et al. [8] harnessed an improved particle swarm optimization algorithm (PSO) alongside a trajectory computation method based on polynomial interpolation. While the enhanced adaptive weights particle swarm exhibits faster convergence and higher accuracy than the standard particle swarm, this method experiences a decline in picking success rates amidst complex environmental conditions surrounding the fruit.
Additionally, path optimization can be facilitated through methods such as neural networks. Wu et al. [116] devised a fusion of Radial Basis Neural Network (RBF) and Denavit-Hartenberg (D-H) modeling to realize trajectory planning for the robotic arm. In their study, D-H modeling was employed for the robotic arm, while the neural network controlled displacement trajectory tracking, allowing for analysis of joint angles and movement speed. Simulation experiments delved into the motion characteristics of linear and circular trajectories, resulting in a 6.84-s reduction in the total motion time of the optimized robotic arm.

e. **Deep Reinforcement Learning (DRL)**
Deep Reinforcement Learning (DRL) is an approach that leverages deep neural networks to enable an agent to learn how to act within an environment to maximize its cumulative reward. In the realm of robotic arm path planning, DRL directs a robot arm's actions by acquiring a policy within a continuous action space. Among the algorithms utilized in this context is Deep Deterministic Policy Gradient (DDPG), which is tailored for learning in continuous action spaces [117]. Lin et al. [118] employed the DDPG algorithm to anticipate collision-free trajectories based on past states. Conversely, certain studies have amalgamated heuristic guidance with deep reinforcement learning algorithms to enhance the operational efficiency and safety of robotic arms. Li et al. [119] fused heuristic guidance with actions derived from the HER-SAC (Soft Actor Critic with Hindsight Experience Replay) algorithm to steer the end-effector toward executing collision-free grasping maneuvers. Lin et al. [6] introduced a Twin Delayed DDPG (TD3) + Automatic Goal Generation (AGG) DPL algorithm to address the inverse kinematics (IK) problem of a series–parallel hybrid banana harvesting robot. This method forecasts the robot's joint values given a goal; however, there remains room for enhancing its localization accuracy.
In addition to conventional path planning methods, several specialized approaches have been implemented. For instance, Wang et al. [120] employed the Coverage Path Planning (CPP) method for kiwifruit picking. CPP converts

the grid-based coverage path planning conundrum into an inter-region Traveling Salesman Problem (TSP), and employs a reinforcement learning algorithm. Typically, CPP is employed for tasks necessitating comprehensive coverage or scanning of an entire area. Within CPP, the robot arm must chart a course ensuring coverage of every point or region within the workspace. This strategy incorporates the principles of reinforcement learning, empowering the robotic arm to learn and optimize the path planning process, thereby enhancing efficiency and performance. Similarly, Park et al. [82] adopted the concept of TSP for job scheduling to mitigate harvest cycle time.

5.3.2 Tasking and Decision-Making Strategies

During the picking process, optimizing system efficiency or meeting particular constraints necessitates sorting, assigning, and arranging tasks to establish the sequence of task execution. This minimizes waiting time and enhances efficiency. In scenarios involving multi-arm or multi-machine collaboration, it's imperative to partition the workspaces of each component to prevent potential conflicts. These conflicts may include instances where two robots converge on the same picking position simultaneously or where one robot's designated work area is encroached upon by another.

Ning et al. [121] employed the Winner-takes-all (WTA) mechanism to plan the harvesting order of bell peppers, drawing from expert experience. Ma et al. [122] utilized a multiple weighting approach to determine the harvesting order of balsam pear in experimental conditions. This approach was compared with the Hybrid Ant Colony-Particle Swarm Algorithm, demonstrating a significant reduction in energy and time consumption. The comparison confirmed the multiple weighting approach's effectiveness in optimizing the harvesting order in terms of rationalization.

Jiang et al. [123] employed a harvesting sequence decision method based on the "sequential mirroring method" utilizing cluster depth information. Additionally, they implemented an asynchronous master–slave dual robotic arm strategy to prevent collisions in hazardous areas. Field performance tests validated that the average harvesting cycle time of the dual-arm robot reached 9 s per bunch, with recognition and harvesting success rates reaching 88 and 83%, respectively. Li et al. [3] utilized a Deep Deterministic Policy Gradient (DDPG) algorithm to determine a sequence of stopping nodes for the robotic vehicle. This approach optimized the robot's operational efficiency by precisely controlling the stopping position and the number of stops. Furthermore, they established a dynamic simulation environment of an industrial greenhouse to enhance the DDPG's generalization ability. This method effectively improved the overall harvesting efficiency of the dual-arm harvesting robot and rationalized task assignments.

Xiong et al. [124] implemented a single-track dual-arm system, where Arm 1 (the primary arm) and Arm 2 (the secondary arm) were distinguished to prevent collisions between the arms. The primary arm was given priority. They proposed a primary and

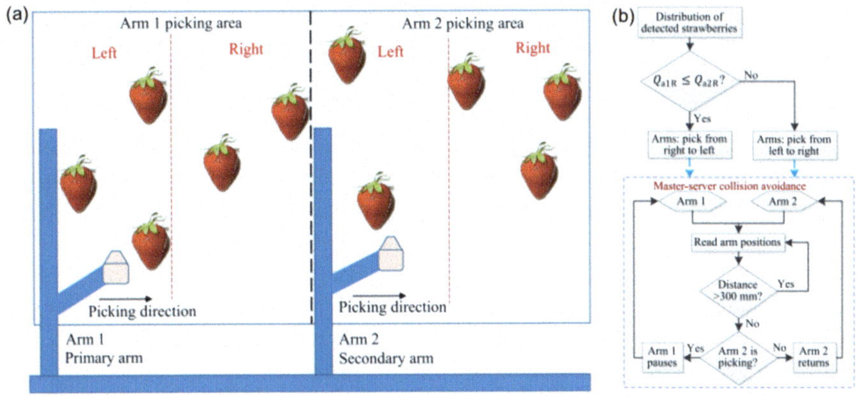

Fig. 5.14 Harvesting order planning [124]

secondary method, as illustrated in Fig. 5.14, to control the arms in a closed loop and avoid collisions.

Barnett et al. [125] developed a task allocation strategy for a multi-armed kiwifruit harvesting robot to ensure equitable fruit distribution. On the other hand, Li et al. [37] introduced a multi-intelligence reinforcement learning (MARL) approach, significantly reducing planning computation time to approximately 1 s. However, operational efficiency in harvesting is hindered by the robot's hardware limitations and the shape of the fruit tree. Additionally, Zhang et al. [126] proposed a multi-arm task allocation method, dividing the robotic arms' workspace using a Monte Carlo method based on random probability. Furthermore, they integrated a simulated annealing algorithm to optimize the picking sequence.

5.4 'Hand': Arm and Hand Design

5.4.1 Robot-Arm Structure

A robotic arm is the core component of an industrial robot and is usually connected by multiple joints, each of which can rotate or move along a specific axis to achieve flexible movement and operation of the arm. A typical robotic arm structure includes a base, joints, linkages, and actuators. For instance, Kang et al. [34] utilized the UR5 industrial robot manipulator manufactured by Universal Robots (UR) in their picking robot, while Wang et al. [13] used the UR3e robot from the same UR family. These robotic arms have six degrees of freedom, enabling flexible movement and manipulation in multiple directions. They feature an easy-to-use programming interface and language, as well as high positioning accuracy. Compared to the UR5, the

UR3e robots are lighter and more compact, making them suitable for tight spaces and delicate operations. They are characterized by flexibility, safety, and ease of use.

In addition to using pre-existing robotic arm products, some studies have focused on designing customized robotic arms based on specific task requirements. For example, Li et al. [12] developed a harvesting system that included a six-degree-of-freedom robotic arm and a pneumatically driven end-effector. Hu et al. [15] designed a configurable four-degree-of-freedom manipulator consisting of four joints: a bottom prismatic joint, a torsional joint, a pitching joint, and another prismatic joint. The working space of the manipulator was calculated and visualized using the Monte Carlo random sampling method. Zhang et al. [127] designed a 4-DOF manipulator with a mechanical structure containing three rotating joints and one moving joint, and the end-effector employed vacuum adsorption, enabling the manipulator to grasp and separate fruits efficiently. Figure 5.15 illustrates the 4-DOF manipulator.

Dual-armed and multi-armed robots have become a trend as robot application scenarios continue to expand and task demands increase. Li et al. [3] utilized two seven-degree-of-freedom xMateER3Pro robotic arms and employed the Monte Carlo method to calculate the dexterous workspace of the robotic arms to meet the demands of complex tasks. Jiang et al. [123] applied a Techsoft TB6-R5 robotic arm, which communicates with the controller in real-time via an EtherCAT bus, allowing the two robotic arms to work independently. These customized robotic arm designs play an important role in different application scenarios and provide more possibilities for the performance and functionality of robotic systems.

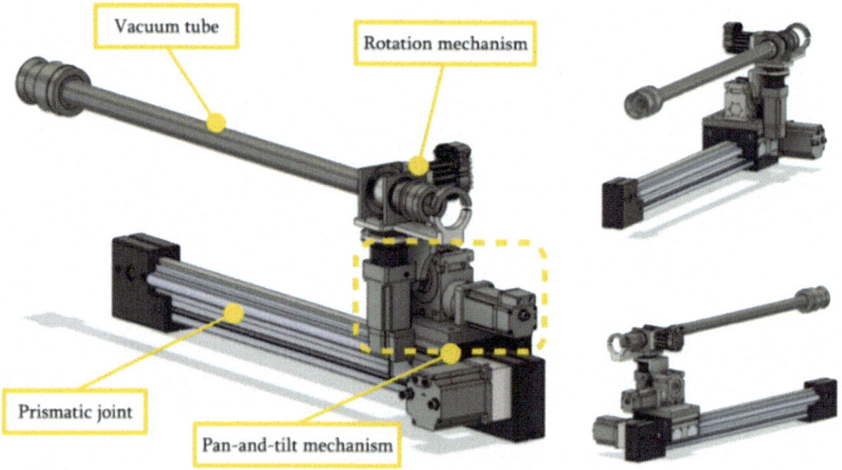

Fig. 5.15 CAD model of the 4-DOF manipulator [127]

5.4.2 *End-Effector*

The end-effector of a picking robot is a crucial device at the end of the robotic arm, responsible for executing the actual crop-picking movements. Various types of end-effectors are commonly used, including cutting, adsorption, and gripping types. Additionally, end-effectors with different structures can be designed according to specific needs. When selecting and designing the end-effector, it is essential to consider the target characteristics, the picking environment, and the requirements of the picking task to ensure the robot can effectively perform the picking task and improve picking efficiency.

5.4.2.1 Cutting Type

A common end-effector design for agricultural robots is the cutter-type module, which is used to sever the fruit from the plant. Jun et al. [128] designed a scissor-type cutting module (Fig. 5.16a), which uses a cutting method to sever tomato fruit stalks. Considering that tomato fruit stalks are relatively hard, the module rotates the winch part and scissor blades through a motor to provide force compensation. After completing the cutting, the scissors return to the initial position by reverse rotation of the motor and contraction of the spring. However, the scissor-type end-effector is prone to conflict in unstructured agricultural environments, reducing the efficiency of harvesting operations. To address this issue, Park et al. [82] designed a circular saw-type module (Fig. 5.16d), which introduces a traction structure to pull the cucumber's stalk to the cutting blade and uses an elliptical suction cup to grip the fruit. Kim et al. [81] developed a cutting module (Fig. 5.16c) that employs a hook-shaped cutter to minimize damage to the fruit surface and features a disk receiver mounted below the cutter to catch the tomato. Arad et al. [129] designed an end-effector (Fig. 5.16b) with a structure for catching the fruit, enhancing the overall efficiency and reducing the likelihood of damage during the harvesting process.

The design of these cutting end-effectors fully considers the structural characteristics of the fruits and the requirements of the operating environment. By employing various cutting methods and auxiliary structures, these end-effectors achieve efficient harvesting of different types of fruits.

5.4.2.2 Gripping Type

Due to the different growing environments and physical properties of fruits, the cutting picking method cannot meet all task requirements. This is why many end-effectors adopt the form of finger gripping to simulate manual picking, providing a more flexible and adaptable picking operation. For instance, in Fig. 5.17a, Gao et al. [130] designed a pneumatic finger-like end-effector using a combination of clamping and rotation for fruit picking in dense environments. This end-effector can

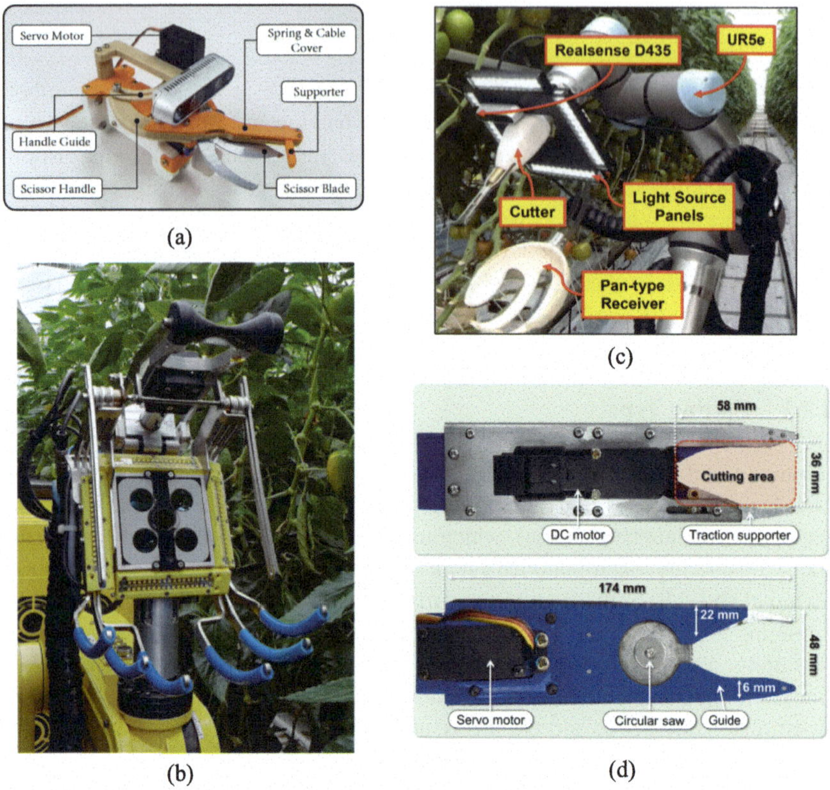

Fig. 5.16 Cutting type end-effectors

perform multiple steps—approach, clamping, separation, removal, and release—with an average picking cycle time of only 6.4 s, a successful separation rate of over 95%, and an average fruit damage rate of only 1.9%. Ma et al. [5] designed an end-effector suitable for kiwifruit picking, consisting of two internally curved grippers, photoelectric sensors, and pneumatic elements, tailored to the shape of the kiwifruit. However, due to the aggregated growth characteristics of kiwifruit, conventional gripper end-effectors may damage neighboring fruits and can only pick one fruit at a time. To address this issue, Fu et al. [131] proposed an envelope-type end-effector, shown in Fig. 5.17b, which allows for the harvesting of kiwifruit in bunches, thereby improving picking efficiency.

Based on the grasping posture parallel to the calyx axis of the apple stem, Fan et al. [132] designed four grasping modes: vertical pulling, horizontal pulling, vertical rotation, and rotation-horizontal pulling. Test results showed that the rotary-horizontal pulling mode under the three-finger grasping posture was the best apple picking mode. Ji et al. [133] designed a flexible three-finger end-effector and applied a circular pull-down separation mode to pick apples. Wang et al. [13] used a pneumatically

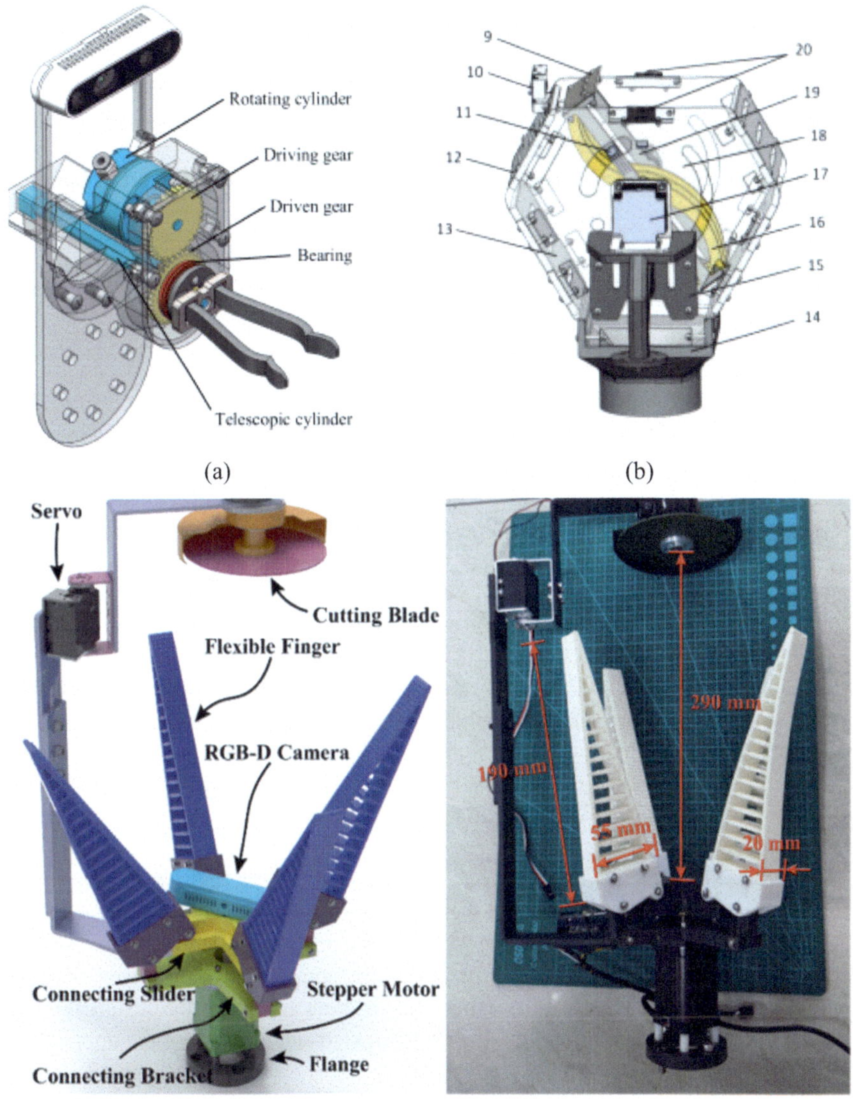

(c)

Fig. 5.17 Gripping type end-effectors

controlled soft gripper for fruit picking. Rong et al. [134] employed a combination of a flexible gripper and a cutting device, shown in Fig. 5.17c. The gripper's finger portion is made of thermoplastic polyurethanes (TPU) material, and the cutting blade is located above the gripper to cut the watermelon stalk. Xiong et al. [124, 135] designed a gripper for strawberry picking, which approaches the fruit from below, pushes the fruit around the target, and swallows the target fruit.

These innovative designs improve picking efficiency while also addressing, to some extent, the adaptability of the end-effector to different fruit morphologies and environments.

5.4.2.3 Adsorption Type

Adsorption end-effectors are also common robotic picking devices used to realize safe and efficient capture and acquisition of target objects. In the field of fruit picking, adsorption end-effectors usually adopt vacuum adsorption technology, capable of accomplishing the picking task without damaging the fruits. Zhang et al. [136] and Hu et al. [15] both used vacuum-based end-effectors for apple picking. Results showed that vacuum adsorption technology can effectively minimize damage to fruits and achieve efficient picking operations. Similarly, negative pressure picking is a common application. In Fig. 5.18a, an adsorption end-effector for mushroom picking is shown. Yang et al. [137] used a suction cup in contact with the mushroom cap, where an air pump pumps the gas inside the suction cup to the external environment, creating a negative pressure source in the flexible chuck. This generates an adsorption force that eventually separates the mushroom from the soil through the rotation of the actuator. The FAR, as shown in Fig. 5.18b, is an autonomous flying robot developed by the Israeli company Tevel Aerobotics, which uses an adsorption structure for its end-effector.

The advantages of adsorption-type end effectors include easy operation, high picking efficiency, and minimal damage to target objects. Therefore, they hold a significant position and promising application prospects in automated picking systems. In the future, with the continuous development and innovation of technology, adsorption end effectors are expected to become more intelligent and flexible, offering greater potential for enhancing performance and expanding the application scope of robot picking systems.

5.5 'Feet': Autonomous Navigation

Autonomous navigation of agricultural robots encompasses pivotal elements like environment sensing, map construction, localization, and path planning. Environment sensing facilitates the acquisition of pertinent information about the surrounding environment, including obstacles, terrain, etc., to furnish data support for subsequent navigation. Map construction entails the conversion of environmental perception

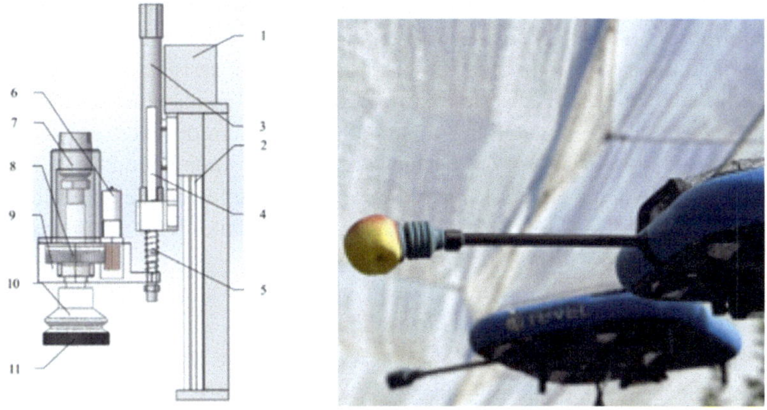

(a) (b)

Fig. 5.18 Adsorption type end-effectors

data into map information, often entailing the generation of two-dimensional or three-dimensional maps, annotated with crucial details such as the orchard layout and the positions of fruit trees. Positioning technologies such as GNSS and IMU are employed to ascertain the precise location of the robot on the map, thereby providing the starting point and endpoint for path planning. During the path planning phase, the robot determines the optimal travel path based on the map information and localization results, taking into account factors such as obstacle avoidance and selection of the shortest route.

Global Navigation Satellite System (GNSS) is a satellite-based navigation system, with the most prevalent system being the Global Positioning System (GPS) developed by the United States. Other notable systems include Russia's GLONASS, the European Union's Galileo, and China's Beidou Navigation Satellite System. Huang et al. [138] devised a voice-controlled auto-navigation picking platform that leverages the BeiDou satellite navigation system for position acquisition and utilizes the Stanley algorithm for path tracking control. Bertoglio et al. [139] engineered a navigation system integrating LiDAR and wheel encoder sensors for in-row, turn-by-turn, and tailgate navigation within row-structured agricultural environments, such as vineyards.

Turning navigation poses a significant challenge in autonomous navigation, demanding advancements in path-planning and motion-planning methodologies for generating turning trajectories, along with corresponding motion control strategies [140]. Gao et al. [141] proposed a headland autonomous turning navigation system for a safflower picking robot, integrating a headland boundary line extraction method, a turning path planning method based on Bezier curves, and a turnaround path tracking and control method. This system exhibits characteristics of high tracking accuracy and robustness. Meanwhile, Mao et al. [14] explored the development of a dual master–slave mode autonomous robot navigation system for orchards. The overall framework of the navigation system is illustrated in Fig. 5.19. In this system, an autonomous navigation orchard transportation robot serves as the master robot, while the navigation orchard picking robot functions as the slave robot. The two robots engage in communication to maintain a minimum distance and avoid collisions. Localization of the robot is achieved through a combination of RTK-GNSS, LIDAR, and odometer. GNSS is utilized to determine target feature points for ground turns, while LIDAR is employed to extract apple trunk features for determining target feature points for row turns.

While GNSS typically operates effectively, signal occlusion or distortion may occur in orchard environments with dense vegetation. Therefore, it is crucial to develop countermeasures for GNSS signal failure. One approach involves fusing multi-sensor data and integrating information from various sensors to enhance positioning accuracy and reliability [142]. Gao et al. [143] investigated a navigation control subsystem based on multi-sensor fusion, employing a wheel odometry + IMU + DGNSS integrated positioning method. Tagarakis et al. [144] utilized an RGB-D camera to map a commercial orchard environment, combining RGB-D information with aerial imagery to achieve high-precision positioning prediction in the orchard. Cerrato et al. [145] proposed a navigation system tailored for structured orchards, which grids aerial images of the target area, integrates sensor data to generate global paths, and utilizes semantic information obtained from segmentation networks to achieve autonomous navigation in the absence of GNSS information. Alternatively, deep reinforcement learning methods can be employed to enhance the robot's navigation capabilities. Martini et al. [146] utilized reinforcement learning to achieve

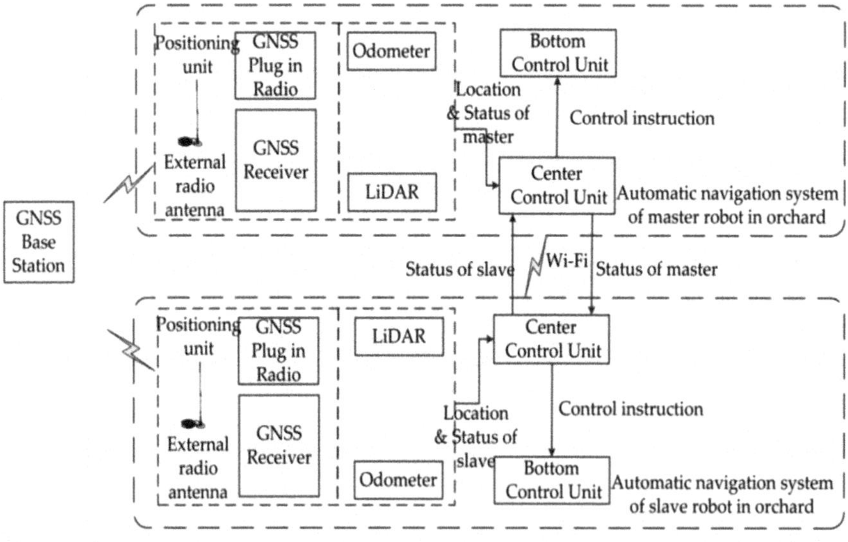

Fig. 5.19 Overall framework of the orchard master–slave navigation system [14]

position-agnostic autonomous navigation of Unmanned Ground Vehicles (UGVs) in a vineyard.

5.6 Challenges and Prospects

Although the research and application of agricultural robots are still in their infancy, the continuous advancement of science and technology, coupled with the development of artificial intelligence, is gradually unveiling broad prospects for the application of intelligent picking robots in agriculture.

5.6.1 Challenges

(1) **Environmental Complexity Challenges**: The complexity of diverse harvesting environments encompasses factors such as irregular terrain, vegetation density, and variations in the growth status of fruit trees. These complexities may induce instability in the robot's movement, thereby limiting sensor observation range and clarity, which can obscure or blur target fruits. Consequently, the robot must possess a high level of environmental adaptability, robustness, and autonomous sensing capability. It should accurately sense and recognize target fruits in complex environments and respond and operate accordingly to the situation.

(2) **Fruit Diversity Challenge**: Due to the diversity of fruits, encompassing variations in shape, size, ripeness, and other factors, designing a general-purpose robot capable of simultaneously adapting to multiple harvesting requirements poses significant challenges. Various types of fruits may exhibit distinct characteristics, such as diverse shapes, ripeness levels, and varying textures, necessitating a high degree of flexibility and adaptability in the robot's sensing, grasping, and handling processes. Consequently, intelligent picking robots must possess the capability to handle different fruit types to accomplish targeted picking and classification effectively.

(3) **Information Processing and Decision Making Capability Challenges**: Picking robots are required to handle substantial sensor data, encompassing environmental information, fruit characteristics, and obstacle locations. Given that sensor data may be susceptible to noise and interference, effective processing of this data and extraction of key information and features are imperative for facilitating rapid and precise decision-making. In intricate and dynamic orchard or farmland settings, robots must possess efficient data processing and computational capabilities, along with robust decision-making skills, to adeptly strategize and execute tasks.

(4) **Mechanical Structure Stability Challenge**: Given the variable and complex operating environment of picking tasks, along with the diverse physical properties of fruits and their distinct requirements, combined with the market's high

demand for product quality, robotic arms must exhibit a high level of flexibility, stability, and accuracy. They must be capable of freely adjusting their attitude and angle to ensure precise fruit picking across various locations. Moreover, the end-effector must accurately grasp and pick fruits based on their characteristics and location, thereby minimizing damage and waste.

(5) **Mobile Platform Design Challenges**: In the horizontal aspect of the environment, terrain irregularities, slopes, and uneven surfaces may result in platform bumps and sliding during movement, thereby compromising stability and safety. Vertically, factors like varying densities and irregular arrangement of fruit trees may lead to inaccurate positioning and loss of satellite signals, necessitating flawless autonomous navigation capability for the mobile platform. Additionally, ensuring continuous operation reliability and efficiency requires the picking robot to possess long-endurance capabilities.

(6) **High Cost**: The development of robots necessitates substantial capital investment and long-term research and development efforts. Additionally, deploying robots requires customization and adaptation for various planting environments, considering factors such as terrain, vegetation, fruit characteristics, etc. Furthermore, the maintenance and upkeep of robots demand significant manpower and material resources, including regular overhauls, maintenance, and troubleshooting work, to ensure their long-term stable operation. Moreover, the picking task is seasonal and cyclical, requiring different resources and manpower investments at different times, thereby further increasing costs and management difficulty.

5.6.2 Prospects

(1) **Industrialization and Large-scale Development of Plantation Industry**: By optimizing the layout and management of orchards, we can improve the growth density and productivity of fruit trees, providing better working environments and operating conditions for picking robots. Applying advanced planting technologies and management modes, such as intelligent greenhouses and soilless cultivation, can improve the quality and yield of fruits while reducing the working difficulty and risks for picking robots. Additionally, optimizing the mechanized facilities and infrastructure of the orchard, including the spacing of planting rows, the shape of the fruit tree canopy, and the drainage system, will facilitate the operation and movement of picking robots. Promoting the transformation and upgrading of agricultural production methods will not only improve the productivity and quality of orchards but also enhance the competitiveness and sustainable development of the entire agricultural industry.

(2) **Development of Sensing Ability and Information Processing Ability**: Sensors will develop in the direction of miniaturization, high precision, and multifunctionality to achieve more accurate and diversified data collection. Multi-sensor fusion technology will be widely used to organically integrate

information from various sensing modalities, such as vision, sound, touch, and odor, to enhance the system's sensing range and accuracy. At the same time, the sensing system will enhance real-time data processing and edge computing capabilities to quickly and efficiently process large amounts of sensor data in complex environments and generate the necessary information and features for real-time decision-making. With the continuous development of intelligent algorithms and networks, the sensing system will become more intelligent and adaptive, capable of automatically learning and optimizing through deep learning, reinforcement learning, and other algorithms. This will allow for more accurate sensing and analysis of complex environments and uncertainties, providing more reliable and accurate support for decision-making.

(3) **Multi-armed and Actuator Modularization Development**: The future robotic arm will develop towards multi-armed and multi-functional capabilities, enabling the intelligent picking robot to possess more powerful picking and processing capabilities. Each robotic arm can operate independently to perform simultaneous picking of multiple fruits or obstacle removal functions. Supported by intelligent algorithms, the multi-armed robot can flexibly coordinate the actions of multiple robotic arms to achieve efficient picking operations, enhancing the robot's applicability and flexibility. Additionally, the modular end-effector design allows the robot to flexibly select and replace actuator modules according to different fruit characteristics and picking requirements. The application of flexible materials is also a development trend for end-effectors. Flexible materials have good deformability and adaptability, which can better accommodate various fruit shapes and characteristics, thereby reducing damage to the fruits.

(4) **Development of Autonomous Navigation and Endurance of the Mobile Platform**: Based on the existing positioning system and multi-sensor fusion technology, the introduction of advanced technologies such as deep learning and neural networks can significantly enhance the intelligent picking robot's understanding of its environment. This enhancement enables the robot to more accurately perceive factors such as the spacing of fruit trees, the morphology of the tree canopy, and the location of obstacles. Utilizing this information, the robot can collaborate with the robotic arm to plan the optimal movement path. This path planning process will minimize idling and repetitive motions of the mobile platform, thereby improving picking efficiency and reducing energy consumption. Additionally, the use of energy-efficient power systems, through the optimization of power management and energy recovery technology, can maximize the working time of the mobile platform, enhancing its continuous operation capability. Future power systems will not only meet the robot's needs for long hours of continuous operation but also help to reduce energy consumption, thereby decreasing dependence on external energy sources and enabling more environmentally friendly and sustainable picking operations.

(5) **Low-cost and Custainable Development**: Through technological innovation and large-scale production, the development of standards for parts and components, and functional modular design, more economical manufacturing

processes and materials can be adopted. This leads to the reduction of production and maintenance costs, and the improvement of component flexibility and applicability. Utilizing remote monitoring and intelligent maintenance technology, the operating status and performance parameters of the robot can be monitored in real time, allowing for timely detection and resolution of issues. Additionally, adopting the sharing economy model, where robots are rented, can reduce users' investment costs and allow for flexible deployment according to the demand for robot resources. This approach improves the utilization rate of the robots and promotes their widespread application, furthering the modernization and scale development of agricultural production.

References

1. Chengliang L, Liang G et al (2022) Current status and development trends of agricultural robots. 53(7)
2. Ji W, Huang X et al (2023) A comprehensive review of the research of the "Eye–Brain–Hand" harvesting system in smart agriculture. Agronomy 13(9)
3. Li Y, Feng Q et al (2024) Intermittent stop-move motion planning for dual-arm tomato harvesting robot in greenhouse based on deep reinforcement learning. Biomimetics (Basel) 9(2)
4. Vrochidou E, Tziridis K et al (2021) An autonomous grape-harvester robot: integrated system architecture. Electronics 10(9)
5. Ma L, He Z et al (2022) A method of grasping detection for kiwifruit harvesting robot based on deep learning. Agronomy 12(12)
6. Lin G, Huang P et al (2022) An inverse kinematics solution for a series-parallel hybrid banana-harvesting robot based on deep reinforcement learning. Agronomy 12(9)
7. Zhang H, Ge Y et al (2022) Picking path planning method of dual rollers type safflower picking robot based on improved ant colony algorithm. Processes 10(6)
8. Gao R, Zhou Q et al (2023) Apple-picking robot picking path planning algorithm based on improved PSO. Electronics 12(8)
9. Xie D, Chen L et al (2022) Actuators and sensors for application in agricultural robots: a review. Machines 10(10)
10. Gai R, Wang X et al (2022) Fruit and vegetable picking robot movement planning: a review. In: 2022 IEEE international conferences on Internet of Things (iThings) and IEEE Green Computing & Communications (GreenCom) and IEEE Cyber, Physical & Social Computing (CPSCom) and IEEE Smart Data (SmartData) and IEEE Congress on Cybermatics (Cybermatics), pp 379–384
11. Aljaafreh A, Elzagzoug EY et al (2023) A real-time olive fruit detection for harvesting robot based on YOLO algorithms. Acta Technologica Agriculturae 26(3):121–132
12. Li J, Tang Y et al (2020) Detection of fruit-bearing branches and localization of litchi clusters for vision-based harvesting robots. IEEE Access 8:117746–117758
13. Wang D, Dong Y et al (2022) Adaptive end-effector pose control for tomato harvesting robots. 40:535–551
14. Mao W, Liu H et al (2022) Development of a combined orchard harvesting robot navigation system. remote sensing. 14(3)
15. Hu G, Chen C et al (2022) Simplified 4-DOF manipulator for rapid robotic apple harvesting. Comput Electron Agric. 199

16. Jiang A, Noguchi R et al (2022) Tree trunk recognition in orchard autonomous operations under different light conditions using a thermal camera and faster R-CNN. Sensors (Basel) 22(5)
17. Kang H, Wang XJC et al (2023) Semantic segmentation of fruits on multi-sensor fused data in natural orchards. 204:107569
18. Redmon J, Divvala SK, Girshick RB, Farhadi A (2015) You Only Look Once: unified, real-time object detection. In: 2016 IEEE conference on computer vision and pattern recognition (CVPR), pp 779–788
19. Wang C-Y, Yeh I-H et al (2024) YOLOv9: learning what you want to learn using programmable gradient information
20. Wang A, Chen H et al (2024) YOLOv10: real-time end-to-end object detection
21. Ren S, He K, Girshick RB, Sun J (2015) Faster R-CNN: towards real-time object detection with region proposal networks. IEEE Trans Pattern Anal Mach Intell 39:1137–1149
22. Liu W, Anguelov D, Erhan D, Szegedy C, Reed SE, Fu C, Berg AC (2015) SSD: single shot multiBox detector. In: European conference on computer vision
23. He K, Gkioxari G, Dollár P, Girshick R (2017) Mask R-CNN. In: 2017 IEEE international conference on computer vision (ICCV), Venice, Italy, pp. 2980–2988. https://doi.org/10.1109/ICCV.2017.322
24. Wang Y, Mu Z (2021) Target recognition and trajectory planning of apple harvesting robot considering color multimedia image segmentation algorithm. Adv Multimedia 2021:1–8
25. Zhu Y, Zhang T et al (2022) Fast location of table grapes picking point based on infrared tube. Inventions 7(1)
26. Ikeda T, Fukuzaki R et al (2021) Tomato recognition for harvesting robots considering overlapping leaves and stems. J Robot Mechatron 33(6):1274–1283
27. Dewi T, Rusdianasari R et al (2022) Image processing application on automatic fruit detection for agriculture industry
28. Luo L, Liu W et al (2021) Grape berry detection and size measurement based on edge image processing and geometric morphology. Machines 9(10)
29. Indra D, Satra R et al (2022) Detection system of strawberry ripeness using k-means. ILKOM Jurnal Ilmiah 14(1):25–31
30. Anraeni S, Indra D et al (2021) Strawberry ripeness identification using feature extraction of RGB and K-nearest neighbour. In: 2021 3rd East Indonesia conference on computer and information technology (EIConCIT), pp 395–398
31. Sekharamantry PK, Melgani F et al (2024) A seamless deep learning approach for apple detection, depth estimation, and tracking using YOLO models enhanced by multi-head attention mechanism. Computers 13(3)
32. Ji W, Pan Y et al (2022) A real-time apple targets detection method for picking robot based on shufflenetV2-YOLOX. Agriculture 12(6)
33. Kang H, Chen CJCEA (2019) Fruit detection, segmentation and 3D visualisation of environments in apple orchards. 171:105302
34. Kang H, Zhou H et al (2020) Real-time fruit recognition and grasping estimation for robotic apple harvesting. Sensors (Basel) 20(19)
35. Yoshida T, Kawahara T et al (2022) Fruit recognition method for a harvesting robot with RGB-D cameras. 9
36. Yoshida T, Onishi Y et al (2022) Automated harvesting by a dual-arm fruit harvesting robot. 9:1–14
37. Li T, Xie F et al (2023) A multi-arm robot system for efficient apple harvesting: perception, task plan and control. Comput Electron Agric 211
38. Zhang G, Tian Y et al (2024) An apple detection and localization method for automated harvesting under adverse light conditions. Agriculture 14(3)
39. Guo J, Yang Y et al (2023) Revolutionizing agriculture: real-time ripe tomato detection with the enhanced tomato-YOLOv7 system. IEEE Access 11:133086–133098
40. Wang Z, Zhu H (2023) Tomato picking robot based on deep learning. J Artif Intell Pract 6:70–73. https://doi.org/10.23977/jaip.2023.060811

41. Mbouembe PLT, Liu G et al (2023) An efficient tomato-detection method based on improved YOLOv4-tiny model in complex environment. Front Plant Sci 14:1150958
42. Zheng T, Jiang M et al (2022) Research on tomato detection in natural environment based on RC-YOLOv4. Comput Electron Agric 198
43. Tsai FT, Nguyen VT et al (2023) Tomato fruit detection using modified yolov5m model with convolutional neural networks. Plants (Basel) 12(17)
44. Yang G, Wang J et al (2023) A lightweight YOLOv8 tomato detection algorithm combining feature enhancement and attention. Agronomy 13(7)
45. Wang X, Wu Z et al (2023) Lightweight SM-YOLOv5 tomato fruit detection algorithm for plant factory. Sensors (Basel) 23(6)
46. Wang C, Wang C et al (2023) A lightweight cherry tomato maturity real-time detection algorithm based on improved YOLOV5n. Agronomy 13(8)
47. Tao Z, Li K et al (2024) Strawberry maturity recognition based on improved YOLOv5. Agronomy 14(3)
48. Luo Q, Wu C et al (2024) A small target strawberry recognition method based on improved YOLOv8n model. IEEE Access 12:14987–14995
49. Cao L, Chen Y et al (2023) Lightweight strawberry instance segmentation on low-power devices for picking robots. Electronics 12(14)
50. Xu Z, Huang X et al (2022) A real-time Zanthoxylum target detection method for an intelligent picking robot under a complex background, based on an improved YOLOv5s architecture. Sensors (Basel) 22(2)
51. Guo J, Xiao X et al (2023) Design and experiment of a visual detection system for zanthoxylum-harvesting robot based on improved YOLOv5 model. Agriculture 13(4)
52. Liu X, Wang D et al (2022) Detection of green asparagus using improved mask R-CNN for automatic harvesting. Sensors (Basel) 22(23)
53. Hong W, Ma Z et al (2023) Detection of green asparagus in complex environments based on the improved YOLOv5 algorithm. Sensors (Basel) 23(3)
54. Wei X, Xie F et al (2023) A study on Shine-Muscat grape detection at maturity based on deep learning. 13
55. Guo C, Zheng S et al (2023) An improved YOLO v4 used for grape detection in unstructured environment. Front Plant Sci 14:1209910
56. Su S, Chen R et al (2022) A novel lightweight grape detection method. Agriculture 12(9)
57. Qiu C, Tian G et al (2022) Grape maturity detection and visual pre-positioning based on improved YOLOv4. Electronics 11(17)
58. Wang W, Shi Y et al (2024) An unstructured orchard grape detection method utilizing YOLOv5s. Agriculture 14(2)
59. Coll-Ribes G, Torres-Rodríguez IJ et al (2023) Accurate detection and depth estimation of table grapes and peduncles for robot harvesting, combining monocular depth estimation and CNN methods. Comput Electron Agric 215
60. Zhou Y, Tang Y et al (2022) Adaptive active positioning of Camellia oleifera fruit picking points: classical image processing and YOLOv7 fusion algorithm. Appl Sci 12(24)
61. López-Barrios JD, Escobedo Cabello JA et al (2023) Green sweet pepper fruit and peduncle detection using mask R-CNN in greenhouses. Appl Sci 13(10)
62. Su L, Sun H et al (2023) Cucumber picking recognition in near-color background based on improved YOLOv5. Agronomy 13(8)
63. Li T, Feng Q et al (2022) Occluded apple fruit detection and localization with a frustum-based point-cloud-processing approach for robotic harvesting. Remote Sens 14(3)
64. Wang C, Li C et al (2023) A performance analysis of a litchi picking robot system for actively removing obstructions, using an artificial intelligence algorithm. Agronomy 13(11)
65. Du X, Meng Z et al (2024) Comprehensive visual information acquisition for tomato picking robot based on multitask convolutional neural network. Biosys Eng 238:51–61
66. Kim W-S, Lee D-H et al (2020) Path detection for autonomous traveling in orchards using patch-based CNN. Comput Electron Agric 175

67. Han Z, Li J et al (2021) Path recognition of orchard visual navigation based on U-Net. 52(1):30–39
68. Song B, Wang Y (2021) Inter-line pose estimation and fruit tree location method for orchard robot. Nongye Jixie Xuebao Trans Chin Soc Agric Mach 52(8)
69. Yang Z, Ouyang L et al (2022) Visual navigation path extraction of orchard hard pavement based on scanning method and neural network. 197:106964
70. Badeka E, Kalampokas T et al (2021) Vision-based Vineyard Trunk detection and its integration into a grapes harvesting robot. Int J Mech Eng Robot Res 374–385
71. Zhou J, Geng S et al (2022) A deep-learning extraction method for orchard visual navigation lines. Agriculture 12(10)
72. Shelhamer E, Long J, Darrell T (2014) Fully convolutional networks for semantic segmentation. In: 2015 IEEE conference on computer vision and pattern recognition (CVPR), pp 3431–3440
73. Lin G, Tang Y et al (2019) Guava detection and pose estimation using a low-cost RGB-D sensor in the field. Sensors (Basel) 19(2)
74. Qi C, Su H, Mo K, Guibas LJ (2016) PointNet: deep learning on point sets for 3D classification and segmentation. In: 2017 IEEE conference on computer vision and pattern recognition (CVPR), pp 77–85
75. Kang H, Wang X et al (2021) Geometry-aware fruit grasping estimation for robotic harvesting in orchards. abs/2112.04363
76. Kang H, Zhou H et al (2020) Visual perception and modeling for autonomous apple harvesting. IEEE Access 8:62151–62163
77. Baisa NL, Al-Diri BJA (2022) Mushrooms detection, localization and 3D pose estimation using RGB-D sensor for robotic-picking applications. abs/2201.02837
78. Retsinas G, Efthymiou N et al (2023) Mushroom detection and three dimensional pose estimation from multi-view point clouds. Sensors (Basel) 23(7)
79. Yin W, Wen H et al (2021) Fruit detection and pose estimation for grape cluster–harvesting robot using binocular imagery based on deep neural networks. 8
80. Wagner N, Kirk R et al (2021) Efficient and robust orientation estimation of strawberries for fruit picking applications. In: 2021 IEEE international conference on robotics and automation (ICRA), pp 13857–13863
81. Kim J, Pyo H et al (2022) Tomato harvesting robotic system based on Deep-ToMaToS: deep learning network using transformation loss for 6D pose estimation of maturity classified tomatoes with side-stem. Comput Electron Agric 201
82. Park Y, Seol J et al (2023) Human-centered approach for an efficient cucumber harvesting robot system: harvest ordering, visual servoing, and end-effector. Comput Electron Agric 212
83. Le Louëdec J, Cielniak GJA (2023) Key point-based orientation estimation of strawberries for robotic fruit picking. abs/2310.11333
84. Zhang F, Gao J et al (2022) Three-dimensional pose detection method based on keypoints detection network for tomato bunch. Comput Electron Agric 195
85. Yoshida T, Fukao T et al (2020) Cutting point detection using a robot with point clouds for tomato harvesting. J Robot Mechat 32(2):437–444
86. Rong Q, Hu C et al (2023) Picking point recognition for ripe tomatoes using semantic segmentation and morphological processing. Comput Electron Agric 210
87. Kim T, Lee D-H et al (2023) 2D pose estimation of multiple tomato fruit-bearing systems for robotic harvesting. Comput Electron Agric 211
88. Bai Y, Mao S et al (2022) Clustered tomato detection and picking point location using machine learning-aided image analysis for automatic robotic harvesting. Precision Agric 24(2):727–743
89. Ci J, Wang X et al (2023) 3D pose estimation of tomato peduncle nodes using deep keypoint detection and point cloud. abs/2311.04699
90. Peng H, Xue C et al (2020) Semantic segmentation of litchi branches using DeepLabV3+ model. IEEE Access 8:164546–164555

91. Qi X, Dong J et al (2022) Method for Identifying Litchi picking position based on YOLOv5 and PSPNet. Remote Sens 14(9)
92. Wang H, Lin Y et al (2022) A study on long-close distance coordination control strategy for litchi picking. Agronomy 12(7)
93. Jin Y, Liu J et al (2022) Far-near combined positioning of picking-point based on depth data features for horizontal-trellis cultivated grape. Comput Electron Agric 194
94. Jin Y, Yu C et al (2022) Detection method for table grape ears and stems based on a far-close-range combined vision system and hand-eye-coordinated picking test. Comput Electron Agric 202
95. Xu Z, Liu J et al (2023) Realtime picking point decision algorithm of trellis grape for high-speed robotic cut-and-catch harvesting. Agronomy 13(6)
96. Zhao R, Zhu Y et al (2022) An end-to-end lightweight model for grape and picking point simultaneous detection. Biosys Eng 223:174–188
97. Zhu Y, Li S et al (2023) Identification of table grapes in the natural environment based on an improved Yolov5 and localization of picking points. Precision Agric 24(4):1333–1354
98. Lu J, Yang Z et al (2023) A machine vision-based method for tea buds segmentation and picking point location used on a cloud platform. Agronomy 13(6)
99. Yan C, Chen Z et al (2022) Tea sprout picking point identification based on improved DeepLabV3+. Agriculture 12(10)
100. Yan L, Wu K et al (2022) Identification and picking point positioning of tender tea shoots based on MR3P-TS model. Front Plant Sci 13:962391
101. Meng J, Wang Y et al (2023) Tea bud and picking point detection based on deep learning. Forests 14(6)
102. Zhang F, Sun H et al (2023) A tea bud segmentation, detection and picking point localization based on the MDY7-3PTB model. Front Plant Sci 14:1199473
103. Zhu L, Zhang Z et al (2023) Detection and localization of tea bud based on improved YOLOv5s and 3D point cloud processing. Agronomy 13(9)
104. Zheng C, Chen P et al (2021) A mango picking vision algorithm on instance segmentation and key point detection from RGB images in an open orchard. Biosys Eng 206:32–54
105. Zhang B, Xia Y et al (2024) Recognition of mango and location of picking point on stem based on a multi-task CNN model named YOLOMS. Precision Agric 25(3):1–23
106. Hu H, Kaizu Y et al (2022) Recognition and localization of strawberries from 3D binocular cameras for a strawberry picking robot using coupled YOLO/Mask R-CNN. Int J Agricult Biol Eng 15(6):175–179
107. Song Z, Zhou Z et al (2021) Canopy segmentation and wire reconstruction for kiwifruit robotic harvesting. Comput Electron Agric 181
108. Zeng S, Hu Y et al (2024) Location of Mulberry leaf picking points based on improved mask RCNN. In: Proceedings of the 2023 4th international conference on artificial intelligence and education (ICAIE 2023), pp 337–344
109. Zhou J, Zhang Y et al (2023) A dragon fruit picking detection method based on YOLOv7 and PSP-Ellipse. Sensors (Basel) 23(8)
110. Cao X, Yan H et al (2021) A multi-objective particle swarm optimization for trajectory planning of fruit picking manipulator
111. Lan J, Xie Y et al (2020) A multi-objective trajectory planning method for collaborative robot
112. Zhang H, Zhu Y et al (2021) Analysis of obstacle avoidance strategy for dual-arm robot based on speed field with improved artificial potential field algorithm. Electronics 10(15)
113. Tian X, Pan B et al (2023) Fruit picking robot arm training solution based on reinforcement learning in digital twin. 11:261–282
114. Ye L, Duan J et al (2021) Collision-free motion planning for the litchi-picking robot. 185:106151
115. Zhang C, Wang H et al (2023) Three-dimensional continuous picking path planning based on ant colony optimization algorithm. PLoS One 18(2):e0282334
116. Wu Z, Du H et al (2022) Artificial intelligence in agricultural picking robot displacement trajectory tracking control algorithm. Wirel Commun Mob Comput 2022:1–12

117. Lindner T, Milecki A et al (2021) Positioning of the robotic arm using different reinforcement learning algorithms. Int J Control Autom Syst 19(4):1661–1676
118. Lin G, Zhu L et al (2021) Collision-free path planning for a guava-harvesting robot based on recurrent deep reinforcement learning. Comput Electron Agric 188
119. Li Y, Feng Q et al (2024) Peduncle collision-free grasping based on deep reinforcement learning for tomato harvesting robot. Comput Electron Agric 216
120. Wang Y, He Z et al (2023) Coverage path planning for kiwifruit picking robots based on deep reinforcement learning. Comput Electron Agric 205
121. Ning Z, Luo L et al (2022) Recognition of sweet peppers and planning the robotic picking sequence in high-density orchards. Comput Electron Agric 196
122. Ma W, Yang Z et al (2023) Study on the fragrant pear-picking sequences based on the multiple weighting method. Agriculture 13(10)
123. Jiang Y, Liu J et al (2022) Development of a dual-arm rapid grape-harvesting robot for horizontal trellis cultivation. Front Plant Sci 13:881904
124. Xiong Y, Ge Y et al (2019) An autonomous strawberry-harvesting robot: design, development, integration, and field evaluation. J Field Robot 37(2):202–224
125. Barnett J, Duke M et al (2020) Work distribution of multiple Cartesian robot arms for kiwifruit harvesting. Comput Electron Agric 169
126. Zhang H, Li X et al (2023) Construction and optimization of a collaborative harvesting system for multiple robotic arms and an end-picker in a trellised pear orchard environment. Agronomy 14(1)
127. Zhang K, Lammers K et al (2023) An automated apple harvesting robot—from system design to field evaluation. J Field Robot. https://doi.org/10.1002/rob.22268
128. Jun J, Kim JK et al (2021) Towards an efficient tomato harvesting robot: 3D perception. Manipul End-Effect 9:17631–17640
129. Arad B, Balendonck J et al (2020) Development of a sweet pepper harvesting robot. J Field Robot 37(6):1027–1039
130. Gao J, Zhang F et al (2022) Development and evaluation of a pneumatic finger-like end-effector for cherry tomato harvesting robot in greenhouse. Comput Electron Agric 197
131. Fu M, Guo S et al (2024) Design and experimentation of multi-fruit envelope-cutting kiwifruit picking robot. Front Plant Sci 15:1338050
132. Fan P, Yan B et al (2021) Three-finger grasp planning and experimental analysis of picking patterns for robotic apple harvesting. Comput Electron Agric 188
133. Ji W, He G et al (2024) A new picking pattern of a flexible three-fingered end-effector for apple harvesting robot. Agriculture 14(1)
134. Rong J, Fu J et al (2022) Development and evaluation of a watermelon-harvesting robot prototype: vision system and end-effector. Agronomy 12(11)
135. Xiong Y, Ge Y et al (2020) An obstacle separation method for robotic picking of fruits in clusters. Comput Electron Agric 175
136. Zhang K, Lammers K et al (2020) System design and control of an apple harvesting robot. abs/2010.11296
137. Yang S, Ji J et al (2022) Modeling and force analysis of a harvesting robot for button mushrooms. IEEE Access 10:78519–78526
138. Huang S, Pan K et al (2023) Design and test of an automatic navigation fruit-picking platform. Agriculture 13(4)
139. Bertoglio R, Carini V et al (2023) A map-free LiDAR-based system for autonomous navigation in vineyards. 2023:1–6
140. Huang P, Zhu L et al (2021) Row end detection and headland turning control for an autonomous banana-picking robot. Machines 9(5)
141. Gao G, Guo H et al (2023) An efficient headland-turning navigation system for a safflower picking robot. J Agric Eng 54(3)
142. Gao P, Lee H et al (2022) Improved position estimation algorithm of agricultural mobile robots based on multisensor fusion and autoencoder neural network. Sensors (Basel) 22(4)

143. Gao G, Guo H et al (2023) Design of a control system for a safflower picking robot and research on multisensor fusion positioning. Engenharia Agrícola 43(2)
144. Tagarakis AC, Filippou E et al (2022) Proposing UGV and UAV systems for 3D mapping of orchard environments. Sensors (Basel) 22(4)
145. Cerrato S, Mazzia V et al (2021) A deep learning driven algorithmic pipeline for autonomous navigation in row-based crops. abs/2112.03816
146. Martini M, Cerrato S et al (2022) Position-agnostic autonomous navigation in vineyards with deep reinforcement learning. 2022:477–484

Chapter 6
Infield Honeysuckle Detection Based on Improved YOLOv5s Under Natural Lighting

Zhaoyu Rui and Zhao Zhang

Abstract Honeysuckle detection through machine vision is an important task for picking honeysuckle timely and efficiently by the picking robot. In this study, image-based object detection methods, including image processing, machine learning (ML), and deep learning (DL), were used to detect honeysuckles in the field to support the vision system of honeysuckle picking robot. A handheld low-cost RGB camera was used to capture 2242 original images containing honeysuckle flowers in different light, angles, and scales. The dataset was expanded to 20,178 using data augmentation where image datasets of 14,747 and 5,431 were used for training and validation, respectively. Three honeysuckle detection models were constructed based on image processing, ML, and DL. The image processing method segments the honeysuckle from the background by color thresholding, hole filling, and corrosion expansion. ML (support vector machine, SVM) and DL (YOLOv3-tiny, YOLOv5s, YOLOv6s, YOLOv7-tiny, and YOLOv8s) models were trained and validated. To solve the difficulties in detection of small targets of honeysuckle, and the imbalance of positive and negative samples, the original YOLOv5s model was improved by fusing the three modules to get an improved YOLOv5s model. The experimental results showed that the overall performance of the improved YOLOv5s model outperformed that of the other methods, with a smaller model size (18.4 MB) and a higher average precision (86.88%). The detection precision was 88.97%, the F_1 value was 84.14%, and the detection speed was more than 30 fps. This study demonstrated that the improved YOLOv5s model can be used as a real-time, reliable, and robust detection method for honeysuckle in the field.

Keywords Image processing · Machine learning · Deep learning · Honeysuckle detection

Z. Rui · Z. Zhang (✉)
Key Laboratory of Smart Agriculture System Integration, Ministry of Education, Beijing 100083, China
e-mail: zhaozhangcau@cau.edu.cn

Key Lab of Agricultural Information Acquisition Technology, Ministry of Agriculture and Rural Affairs, China Agricultural University, Beijing, China

Y. Yun et al. (eds.), *Advanced Sensing and Robotics Technologies in Smart Agriculture*, Smart Agriculture 9, https://doi.org/10.1007/978-981-97-6441-9_6

6.1 Introduction

Honeysuckle is a semi-evergreen twining vine that is widely planted. Its petals are oval or elliptical. It has the effects of clearing heat, detoxifying, reducing inflammation, and reducing swelling [1]. It can be used as a traditional Chinese medicine and made into tea. It has an extremely high economic value. During the flowering stage, honeysuckle only takes a few days to grow from buds to bloom (2). Once it blooms, the economic value is greatly reduced, so it must be harvested quickly. Currently, the main picking methods for honeysuckle are manual picking, or small hand-held pickers and backpack pickers. Manual picking is labor-intensive, low-efficiency, and high-cost. The small handheld picking machine uses a pair of relatively rotating rubber rollers to clamp and suck the honeysuckle. Although it avoids picking by hand and improves efficiency, the capacity of the collection bin is small. When the bin is full, the harvested honeysuckle needs to be poured out, which greatly reduces the picking efficiency. The backpack picker adds a larger backpack collection bin on the basis of the handheld picker, which effectively increases the load, but also increases the work intensity of the staff and easily causes labor injuries. For the above picking methods, honeysuckle harvesting requires the employment of a large number of laborers for a long time, which not only results in extremely high labor costs, but also faces the problem of labor shortage. In addition, the placement of large numbers of people is also a thorny issue [3]. To solve these issues, it is necessary to develop an efficient honeysuckle picking robot to replace manual work. The honeysuckle picking robot mainly consists of a vision system and picking mechanical equipment. The primary task of the picking robot is to detect honeysuckle. The accuracy of honeysuckle detection of the vision system directly affects the honeysuckle picking effect.

With the development of machine vision technology, many researchers have applied traditional image processing methods for crop target detection [4]. Traditional image processing methods generally extract objects from images based on features such as shape, color, or texture. Zheng [5] used the color space model method combined with morphological operations to perform a threshold segmentation operation on the honeysuckle to segment the honeysuckle from the background, and then used the edge detection canny operator to detect the honeysuckle. The honeysuckle recognition rates of the threshold segmentation method and the canny operator are 79.17% and 66.67%, respectively.

With the escalation of machine learning (ML), researchers use ML algorithms such as support vector machines (SVM), decision trees (DT), K nearest neighbor (KNN), random forest (RF), and neural networks (NN) for improving the accuracy and robustness of crop detection [6]. [7] used methods such as frequency filtering, segmentation, and feature extraction, combined with different ML classification techniques such as discriminant analysis (DA), KNN, and SVM, to classify wheat ears in RGB images with an accuracy of more than 90%. [8] collected color RGB and color infrared (CIR) images of volunteer corn and soybean seedlings, extracted 20 color, shape, and texture-related features, and input the datasets into different ML classifiers: KNN, SVM, NN, and RF. SVM has the highest detection accuracy for volunteer corn, which is close to 100%. The performance of an ML model depends largely on whether the features chosen by the researcher are representative of the plant's characteristics.

[8] collected color RGB and color infrared (CIR) images of volunteer corn and soybean seedlings, extracted 20 color, shape and texture related features, and fed the dataset into different ML classifiers: KNN, SVM, NN and RF, SVM has the highest detection accuracy for volunteer corn, which is close to 100%. The performance of an ML model depends largely on whether the features chosen by the researcher are representative of the plant's characteristics.

Deep learning (DL) algorithms represented by convolutional neural networks (CNN) contain more hidden layers and can extract more abstract features without artificial features 9. Therefore, traditional image detection algorithms are gradually replaced by DL algorithms. Commonly used target detection algorithms include two-stage and single-stage detection algorithms. Since the two-stage detection algorithm achieves target detection by generating proposed boxes, it has high detection accuracy but slow detection speed. It mainly includes Faster R-CNN [10], Cascade R-CNN [11], and SPPnet [12]. The single-stage detection algorithm directly detects the target without generating candidate frames and is faster but has lower accuracy than the two-stage algorithm. It mainly includes SSD [13], RetinaNet and YOLO [14, 15] series algorithms. Among the single-stage detection algorithms, the YOLO series algorithms have been widely used in agriculture because they can ensure detection accuracy at a faster detection speed. The YOLOv4 [16] algorithm based on channel pruning was used to achieve real-time and rapid detection of apple inflorescences [17]. They discussed the detection performance of apple inflorescences under different tree species and lighting conditions. A fast and accurate target detection algorithm by improving YOLO-tiny [18] was developed to achieve rapid detection of kiwi fruit in orchards [19]. Pre- and post-processing technology using YOLOv3 was proposed and applied it to an apple harvesting robot machine vision system [20]. Its average detection time and error rate were lower than all known similar systems. Based on the YOLOv5s DL algorithm for channel pruning, [21] developed an accurate detection method for small apple fruits in a miniature model size, which can provide a reference for the development of portable mobile fruit pruning terminals. The YOLOv5s model was further improved by fusing pyramid split attention (PSA) and Bi-directional feature pyramid network (BiFPN) and combining them with the K-means algorithm [22]. The model could detect spikelets that need to be removed from grape inflorescences, with an accuracy of 78%.

There are few studies on the detection of honeysuckle in the field since most of the research focuses on the detection of other crops. Therefore, the purpose of this study is to evaluate the potential of machine vision technology for direct detection of honeysuckle in the field, and to provide the necessary technical support for the honeysuckle picking robot vision system to realize honeysuckle detection. The specific objectives of this study were to: (i) construct a honeysuckle dataset through infield image collection, labeling, and augmentation; (ii) develop image processing,

ML, and DL models to detect honeysuckle; (iii) compare the performance of three methods in field honeysuckle detection and recommend the most satisfactory method.

6.2 Materials and Methods

For honeysuckle detection, three steps were followed in this research (Fig. 6.1). The steps were: (i) Data set construction by collecting honeysuckle RGB images from the field, (ii) Honeysuckle detection using image processing technology and building ML and DL object detection models, and (iii) Compare and evaluate the performance of different detection methods. Brief details of each step in terms of methods and parameters are shown in Fig. 6.1.

6.2.1 Data Acquisition and Processing

6.2.1.1 Data Acquisition

The honeysuckle data collection location is located at the Honeysuckle Plantation in Julu County, Hebei Province, China. The data was collected from August 18 to August 19, 2023. The variety of honeysuckle was Juhua-1. The shooting time was from 7:00 to 18:00. The shooting equipment was a Nikon D5300 camera. The shooting was taken place under natural light in various conditions, including close range, long distance, vertical canopy, sloping canopy, light and dark, etc. The shooting distance was from 20 to 100 cm. A total of 2383 photos were captured, with the picture resolution of 2464 $\times$ 1632. A data collection of 2242 original honeysuckle photos were obtained after removing blurry photographs, as shown in Fig. 6.2.

6.2.1.2 Image Labeling

Honeysuckle has four forms: normal honeysuckle buds (Fig. 6.3a), curled honeysuckle buds (Fig. 6.3b), bloomed honeysuckle flowers (Fig. 6.3c), and tiny honeysuckle buds (Fig. 6.3d). As shown in Fig. 6.3, normal honeysuckle buds have the highest economic value. Although curled honeysuckle buds and bloomed honeysuckle flowers have lower economic values, they still have a demand for picking. Therefore, the three forms of honeysuckle, including normal honeysuckle buds, curled honeysuckle buds, and bloomed honeysuckle flowers are all labeled as "honeysuckle". The tiny honeysuckle buds do not meet the picking requirements, so they are not marked. Labelimg software was used for annotation. A labeled dataset of 2,242 images was obtained, with a total of 46,872 honeysuckles labeled, and each picture contained approximately 21 honeysuckles on average.

Fig. 6.2 Honeysuckle data collection. (**a**) Image acquisition under different distance, angle and light conditions; (**b**) honeysuckle image examples under different conditions, (I) long-distance shooting, (II) close-up shooting, (III) vertical shooting, (IV) oblique shooting, (V) bright light, (VI) dark light

6.2.2 Different Methods for Honeysuckle Detection

6.2.2.1 Image Processing

Image processing methods usually extract objects based on features such as color, texture, and shape. The challenges in extracting images of honeysuckle that: (i) the shapes of honeysuckle vary greatly when photographed taken from different

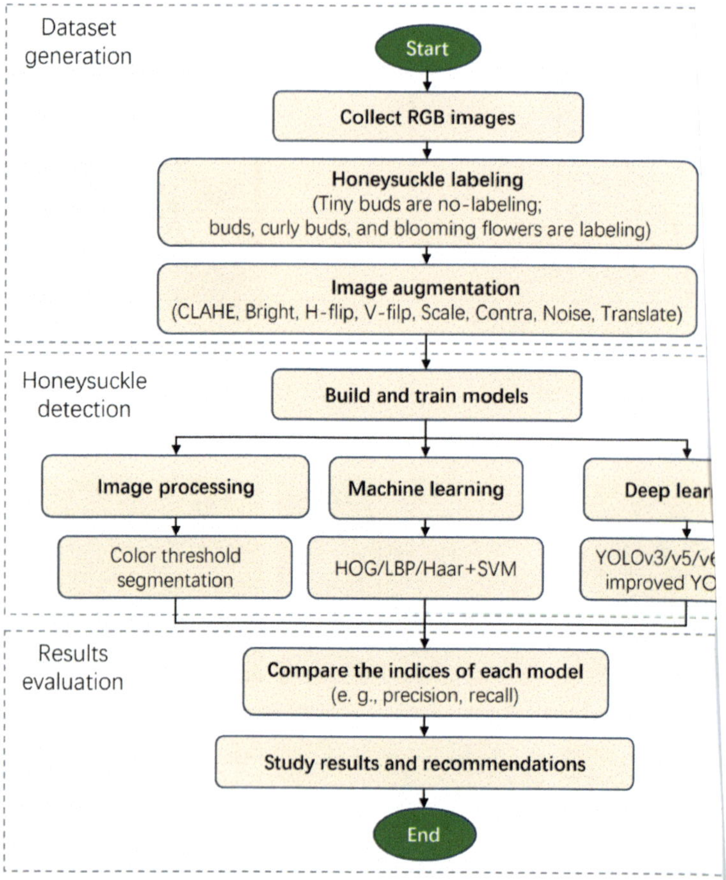

Fig. 6.1 Overall process flowchart of infield honeysuckle detection using image machine, and deep learning. CLAHE, contrast limited adaptive histogram equalization horizontal; Contra, random contrast, V-flip, flip vertical; HOG, histograms of oriente LBP, local binary patterns; Haar, Haar wavelet; SVM, support vector machine

6.2.1.3 Image Augmentation

In order to enrich the dataset, prevent overfitting, and improve the robustn model, eight methods were used to enhance the image data: CLAHE (contra adaptive histogram equalization) processing, brightness adjustment, flip h scale variation, random contrast, flip vertical, add noise, and translation, as Fig. 6.4. The dataset was expanded to 20,178 images.

module in a specific direction, such as horizontal, vertical, or diagonal). HOG, LBP, and Haar features focus on the structure, shape, and texture of objects and the changes in light and dark of objects. These features are reasonably assumed to be honeysuckle features.

Input the extracted HOG, LBP, and Haar features into the SVM classifier for training; use the trained classifier to detect features in the sliding window area and classify them as positive or negative. After the judgment is completed, move to the next window and complete the picture. After the traversal, the NMS algorithm with an IoU (intersection over union) threshold of 0.5 is used to merge the target frames, remove duplicate frames, and finally complete the detection of honeysuckle.

6.2.2.3 Deep Learning

Object detection based on DL can be divided into two stages (extracting ROIs and then refining the location and category of the target) and one stage (using a single network to predict the category and location coordinates of the target directly). According to existing research [26–28], the single-stage method has more advantages regarding real-time accuracy and target detection. YOLO is a single-stage target detection algorithm based on regression. As a superior-performance version of the YOLO series, YOLOv5 integrates various optimization strategies to balance speed and accuracy. YOLOv5 is divided into x, l, m, and s versions according to the width and depth of the model (https://github.com/ultralytics/yolov5). YOLOv5s is the network with the most minor network depth and feature map width, and the other three networks are deepened and widened based on it. Therefore, YOLOv5s was chosen as the fundamental core network in this investigation, and the enhancement techniques were further examined.

Improved YOLOv5s

YOLOv5s consists of input, backbone, neck, and prediction. The backbone network is mainly used for image feature extraction. The backbone network consists of four modules: focus, bottleneck, C3, and SPP [29]. Conv is the smallest component of the YOLOv5 residual network block, consisting of a convolution layer, a batch normalization (BN) layer, and a SiLU activation function. C3 is the most important residual feature learning module in YOLOv5. Its structure is divided into two branches: one overlays the module with three ordinary convolutional layers with multiple bottlenecks, and one branch passes through one ordinary convolutional layer. Finally, the two branches are connected. The neck network adopts the FPN + PAN structure, which enables the model to obtain rich feature information [30, 31]. The detection layer predicts image features through output, generates bounding boxes, predicts categories, and records defects' location and classification information. Next, the final prediction output is executed.

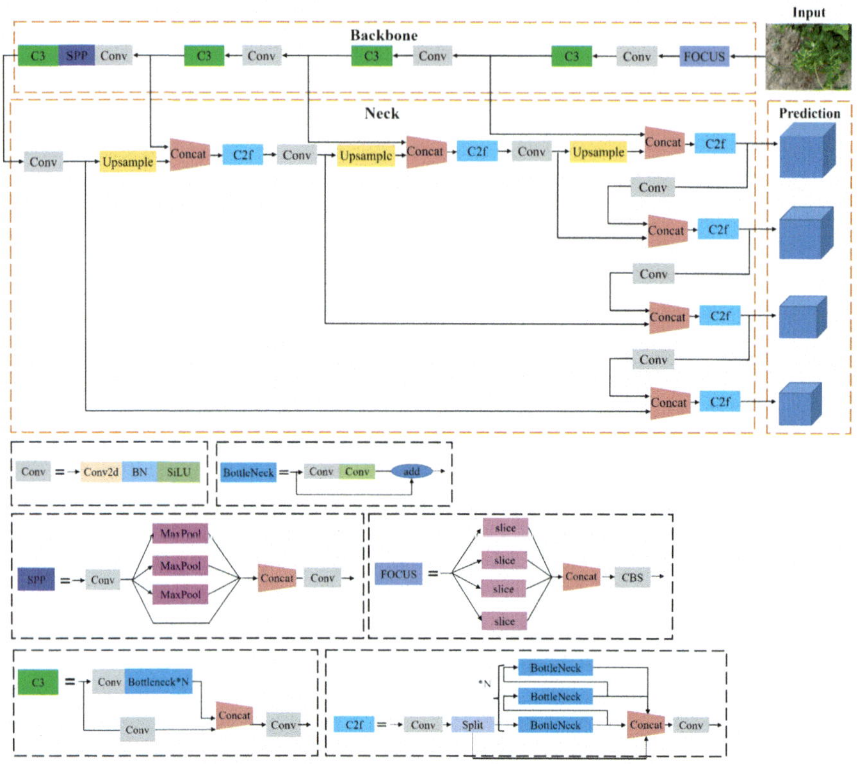

Fig. 6.6 Improved YOLOv5s model structure

In order to further improve the detection performance of the YOLOv5s model, this study proposed an improved YOLOv5s honeysuckle detection network based on the original model, as shown in Fig. 6.6. The neck and prediction components of the network were enhanced to address the issues of difficulties in recognizing tiny honeysuckle targets and the imbalance of positive and negative data. The model improvement was mainly completed by adding or replacing the following three modules: (i) tiny object detection layer; (ii) C2f module; and (iii) Focal-EIoU loss.

Tiny object detection layer

Statistical analysis was performed on all labeled boxes in the honeysuckle dataset. According to the definition of target size in the COCO (common objects in context) dataset [32], the targets were classified into three categories: small size targets with a total number of pixels less than 32×32,medium-size target with the total number of pixels between 32×32 and 96×96; and a large-size target with a total number of pixels greater than 96×96. As shown in Fig. 6.7, there are 329,897 annotation boxes in the Honeysuckle dataset, including 273,283 small-sized targets, accounting for 82.8% of the entire data set, 55,585 medium-sized targets, accounting for 16.9%, and large-sized targets only accounting for 0.03%.

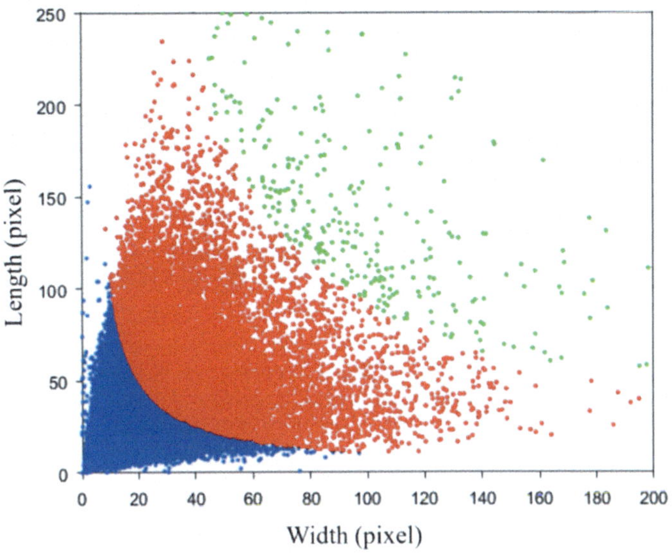

Fig. 6.7 Size distribution of honeysuckle annotation boxes. Red indicates a small target, blue indicates a medium target, and green indicates a large target

In this dataset, the honeysuckle to be detected in most RGB images are small targets randomly distributed in different locations. To see them accurately, it is necessary to extract both the edge information of each honeysuckle and the global features of all images. One reason YOLOv5 does not have a good detection effect on small targets is that the size of small target samples is small, and the down-sampling multiple of YOLOv5 is relatively large [33]. It is challenging to learn the feature information of small targets with deeper feature maps, so it is proposed to add tiny object detection layer. The detection layer detects the shallow and deep features maps after splicing them. Adding a small target detection layer can make the network pay more attention to detecting small targets and improve the detection effect.

C2f module

As shown in Fig. 6.6, the C3 structure in YOLOv5 includes three standard convolutional layers and N bottleneck modules. This module is the main module for learning residual features. Its structure is divided into two branches. One uses the multiple bottleneck stacks and three standard convolution layers specified above. The other only goes through a basic convolution module. Finally, perform the concat operation on the two branches. The C3 module adopts a bottleneck structure, which allowing the C3 module to better capture context information and perform better when detecting larger targets. The C2f module extracts features through a series of convolutional and pooling layers and enlarges the feature map to the same size as the original input image through upsampling operations. It can directly connect lower-level to higher-level feature maps, effectively preserving more spatial information and giving it

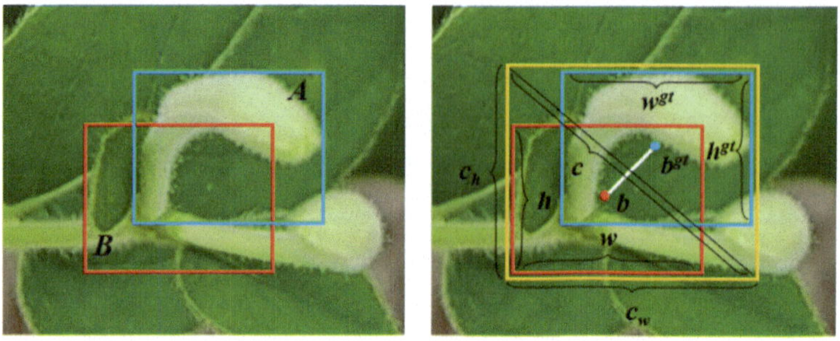

(a) (b)

Fig. 6.8 Quantitative comparison of IoU and EIoU. *A*, ground true; *B*, prediction; *b*, *w*, *h* and b_{gt}, w_{gt}, h_{gt} are the center, width, and height of the prediction and ground truth; *c* is the diagonal length of the minimum external rectangle of the two bounding boxes, and c_w and c_h are the width and height of this rectangle, respectively

better performance when dealing with smaller targets. Replacing the C3 structure of YOLOv5 with the C2f structure not only ensures lightweight while obtaining richer gradient flow information, but also adjusts different channel numbers for other scale models. In addition, due to the smaller convolution kernel size of the C2f module, it is more sensitive to recognizing small targets.

Focal EIoU loss

YOLOv5 defaults to using CIoU loss as a function, but CIoU cannot reflect the actual difference between width and height and their confidence, which sometimes hinders the model from effectively optimizing similarity. To solve the problem of CIoU, researchers proposed EIOU loss [9] based on the aspect ratio impact factor decomposition of CIOU which was used in this study. The loss function consists of three parts: IoU loss (L_{IoU}), center distance loss (L_{dis}), and aspect ratio loss (L_{asp}). EIoU minimizes the difference in width and height between the actual box and the predicted box, thereby accelerating convergence speed and improving regression accuracy. The IoU loss and EIoU loss of bounding box regression are shown in Fig. 6.8.

The calculation formulas are as follows:

$$IoU = \frac{A \cap B}{A \cup B} \tag{6.1}$$

$$L_{EIoU} = L_{IoU} + L_{dis} + L_{asp} = 1 - IoU + \frac{\rho^2(b, b^{gt})}{c^2} + \frac{\rho^2(w, w^{gt})}{c_w^2} + \frac{\rho^2(h, h^{gt})}{c_h^2} \tag{6.2}$$

Where A is the ground truth box, B is the prediction box, $\rho^2(*,*)$ denotes the Euclidean distance function, (b, w, h) and (b^{gt}, w^{gt}, h^{gt}) are the center, width, and height of the prediction and ground truth, c is the diagonal length of the minimum external rectangle of the two bounding boxes, and c_w and C_h are the width and height of this rectangle, respectively. However, the honeysuckle regions are unregular minor objects, which results in most of the predictions being low-quality (i.e., the bounding boxes with very low EIoU) during training.

To avoid the deterioration of training results caused by this sample imbalance, Focal-EIoU was further introduced in the loss function to balance the weights of high-quality bounding boxes and low-quality bounding boxes and reduce a large number of anchor box pairs that had less overlap with the target box. The optimization contribution of bound box regression enabled the regression process to focus on high-quality anchor boxes. The calculation formula for the loss function was:

$$L_{Focal-EIoU} = IoU^{\gamma}(1 - EIoU) \tag{6.3}$$

where γ is the parameter that controls the weight. In the above equation, IoU is employed as a quality indicator.

Specifically, for the low-quality bounding box with a small IoU, its weight IoUγ will be smaller, aiming to reduce the loss from the low quality; on the contrary, for the high-quality bounding box with a large IoU, its weight IoUγ will be larger, to supply more high-quality loss. By doing so, Focal-EIoU can ensure that high quality bounding boxes provide more contributions to the optimization, avoiding the collapse of the model and accelerating convergence.

Model Training

Model training was performed in a 64-bit Windows system using an NVIDIA GeForce GTX 1080Ti GPU with 11 GB of memory. The Python 3.8 as the programming language and Pytorch 13.0 as the deep learning framework were selected and used in this study. The collected honeysuckle images were used as experimental data, and the input size was normalized to 640 × 640 × 3 and divided into a training set and a verification was set according to the ratio of 8:2. The batch size was set to 8. The learning rate adaptive strategy was adopted to speed up the convergence of the model and was ensured its stability. The initial learning rate was 0.01, and the weight decay coefficient was 0.0005. The maximum value of the epoch was set to 100.

6.2.3 Evaluation Metrics

Several widely used assessment metrics were chosen to assess the model's performance quantitatively across several dimensions: average precision (AP), F_1, precision (P), and recall (R). Furthermore, considering the lightweight requirements for model deployment, model size was also chosen as an assessment indicator [34], The evaluation index formulas are as follows:

$$P = \frac{TP}{TP + FP} \tag{6.4}$$

$$R = \frac{TP}{TP + FN} \tag{6.5}$$

$$AP = \int_0^1 P(R)dR \tag{6.6}$$

$$F_1 = \frac{2 \times P \times R}{P + R} \times 100\% \tag{6.7}$$

6.3 Results and Discussion

6.3.1 *Honeysuckle Detection Through Image Processing*

The honeysuckle detection results based on color threshold segmentation and morphological processing methods show that the average P and R values of this method are 55.80 and 65.47%, respectively (Appendix_1). The reasons for the low P and R may be the following two points: Under strong light conditions, the brightness of the stems, leaves, and exposed areas is similar to that of honeysuckle, and only the color threshold is used to segment them, thus causing false detections. The accuracy is low; the brightness of some immature honeysuckles that appear green is reduced, and it is easy to remove them as backgrounds mistakenly; in addition, it is easy to identify two or more honeysuckles as one, and the recognition effect is poor. The segmentation method based on color features has better recognition results for honeysuckle shots at close range. Still, it requires manual threshold adjustment to adapt to honeysuckle images of different scales. It is susceptible to light and has a high recognition error rate, making it difficult to meet the requirements of actual picking operations.

6.3.2 *Results of Machine Learning Based Method*

Following training, the classification accuracy of HOG, Haar, and LBP feature with SVM classifiers was found 90.67%, 90.43%, and 62.70%, respectively. According to the categorization findings based on three ML classifiers, the HOG + SVM classifier had the highest precision during classification. This may be because the HOG feature uses gradient or edge-direction density distribution to describe the appearance and shape of the object. There are obvious shape differences between the honeysuckle and the background in the image, so the HOG feature values of honeysuckle are

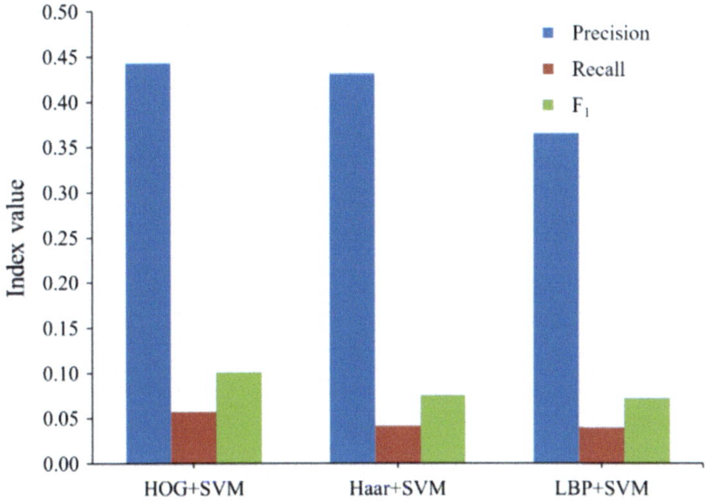

Fig. 6.9 Machine learning based honeysuckle detection results

easier to extract and use for classification and detection. The accuracy of the Haar + SVM detector is slightly lower than that of HOG, probably because Haar uses comprehensive features (horizontal, vertical, and diagonal directions) to express the gradient changes of local pixels and has good sensitivity to edges and lines with obvious color changes, but the color difference between the honeysuckle and the background is small. The LBP + SVM detector has the lowest accuracy because LBP is an operator used to describe the local texture features of an image, and the texture of honeysuckle and stems are similar, making it difficult to distinguish. Therefore, it can be concluded from the above analysis that the HOG + SVM classifier has the best performance among the three methods.

Applying the above three methods to the detection of honeysuckle in images, the P values based on HOG + SVM, Haar + SVM, and LBP + SVM detectors are 44.43%, 43.13%, and 36.48%, respectively (Fig. 6.9). Compared with classification The accuracy has dropped significantly, possibly because the scales of the honeysuckle in the picture are different, making larger and smaller honeysuckles difficult to detect or mis-detected. Therefore, the above ML-based method is not robust to scale changes.

6.3.3 Deep Learning Based Method

6.3.3.1 Model Training Process

Figure 6.10 shows the training loss, AP and PR curve of improved YOLOv5s compared to the original YOLOv5s. As the number of training rounds increases, both

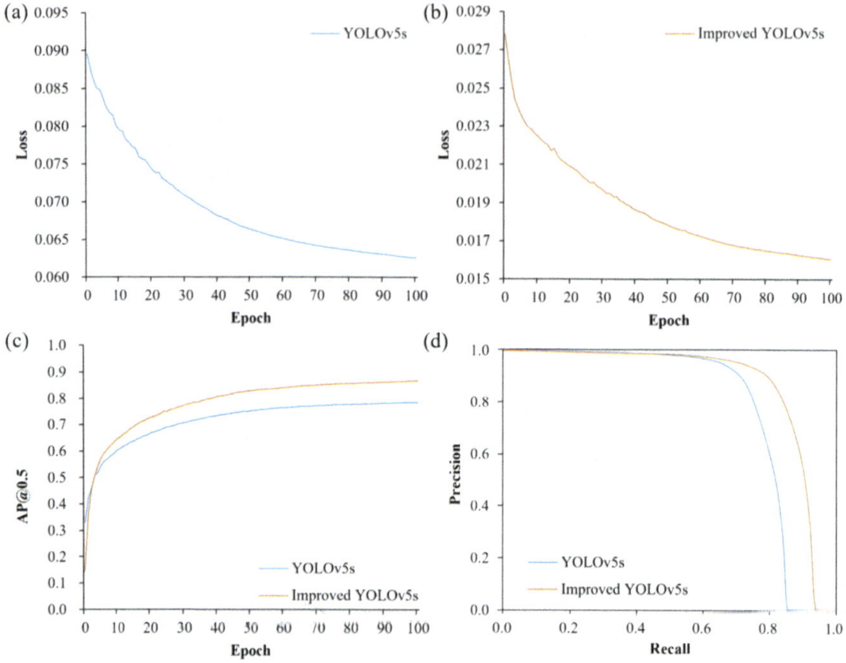

Fig. 6.10 Comparison of training results between YOLOv5s and improved YOLOv5s models. (**a**) Loss curve of YOLOv5s, (**b**) loss curve of improved YOLOv5s, (**c**) AP@0.5 curve, and (**d**) PR curve

losses show a downward trend, while AP shows an upward trend. Compared with the loss of the original YOLOv5s, the loss of improved YOLOv5s was significantly reduced. The AP has been improved, indicating that improved YOLOv5s, which combined three modules, has more vital fitting and generalization capabilities.

6.3.3.2 Ablation Experiments

As improved YOLOv5s with three modules (tiny object detection layer, C2f module, and Focal-EIoU) was proposed to solve the problem of difficulty in honeysuckle small object detection and imbalance of positive and negative samples. Seven variations of improved YOLOv5s were created by carefully combining several modules to confirm the viability and efficacy of each module. For the correctness of the ablation experiments, the model hyperparameter settings and runtime environment are the same. The ablation experiments and experimental results are detailed below to clarify the impact of different modules on the final performance.

Table 6.1 shows the ablation test results of these variants on the validation set. The results showed that tiny object detection layer had the most significant impact, increasing AP to 85.57% and F_1 to 83.17%. After integrating C2f and tiny object

Table 6.1 Ablation experiments result of different models. A is a tiny object detection layer; B is a C2f module; C is Focal-EIoU

Model				P/(%)	R/(%)	AP@0.5/(%)	Model size/(MB)	F_1/(%)
Number	A	B	C					
(1)				89.68	70.69	78.56	13.7	79.06
(2)	√			89.11	77.98	85.57	15.1	83.17
(3)		√		88.42	70.09	77.61	18.0	78.20
(4)			√	88.06	72.63	79.83	13.7	79.60
(5)	√	√		89.49	79.45	86.31	18.4	84.17
(6)	√		√	88.34	78.52	86.06	15.1	83.14
(7)		√	√	87.73	71.74	79.06	18.0	78.93
(8)	√	√	√	88.97	79.80	**86.88**	18.4	84.14

detection layer, compared with using the tiny object detection layer alone, although the model size increased slightly (3.3 MB), P, R, AP and F_1 increased by 0.38, 1.47, 0.74 and 1.00% respectively. Although the P value of improved YOLOv5s was slightly lower than the original model (reduced by 0.71%). However, the R, AP, and F_1 of the improved model were improved by 9.11, 8.32, and 5.08%, respectively, compared with the original model. The model size is also within the acceptable range.

6.3.3.3 Comparison of Different Detection Algorithms

This study performed comparison studies using YOLOv3-tiny, YOLOv6s, YOLOv7-tiny, and YOLOv8s to confirm the detection effect of upgraded YOLOv5s. The input size of each baseline was standardized to 640 × 640 × 3, and the hyperparameters were left unaltered to guarantee the fairness of the comparison analysis. The baselines were trained for 100 epochs, and the optimal trained model for each baseline was selected. The experimental results are shown in Table 6.2. The results show improved YOLOv5s have the best AP@0.5, R, and F_1 values. The AP@0.5 of improved YOLOv5s is 33.51, 8.32, 7.08, 15.02, and 7.35 percentage points higher than YOLOv3-tiny, YOLOv5s, YOLOv6s, YOLOv7-tiny, and YOLOv8s, respectively. The model size of Improved YOLOv5s is only slightly higher than YOLOv5s and YOLOv7-tiny but still within the acceptable range.

The honeysuckle detection performances of improved YOLOv5s with other original models are shown in Fig. 6.11. From the detection results, it can be seen that for different scales of honeysuckle, improved YOLOv5s could correctly detect other original models that missed or missed detections of honeysuckle, identify partially occluded honeysuckle, and had higher confidence than original model. The prediction box could fully envelop the honeysuckle and identify overlapping honeysuckles. The comparison results showed that the three proposed modules could effectively

Table 6.2 Comparing the performance of YOLO series detectors on honeysuckle dataset

Model	P/(%)	R/(%)	AP@0.5/(%)	Model size/(MB)	F_1/(%)
YOLOv3-tiny	76.15	46.47	53.37	24.4	57.72
YOLOv5s	89.68	70.69	78.56	13.7	79.06
YOLOv6s	88.50	72.90	79.80	38.7	79.95
YOLOv7-tiny	76.95	66.09	71.86	11.7	71.11
YOLOv8s	85.64	70.80	79.53	21.4	77.52
Improved YOLOv5s	88.97	79.80	**86.88**	18.4	84.14

Fig. 6.11 Comparison of detection results of the YOLOv3-tiny, YOLOv5s, YOLOv6s, YOLOv7-tiny, YOLOv8s, and improved YOLOv5s models. The yellow square box indicates a missed detection

improve the model detection performance, the AP of the improved yolov5s reached 86.88%, indicating that the constructed network is reliable and robust.

6.3.4 Comprehensive Comparison of Three Methods

According to the analysis of the detection results of the above three methods (image processing, ML, and DL), it can be known that color threshold segmentation, HOG + SVM detector, and improved YOLOv5s are the optimal models for each method. The detection accuracy P of the color threshold segmentation and HOG + SVM detectors is very low (65.47%, 44.43%), which makes it difficult to meet the needs of practical applications. However, the P of improved YOLOv5s exceeds 85%, and all indicators are the best among the three methods.

The above analysis shows that the overall detection performance of the color threshold segmentation and HOG + SVM detectors is much lower than the improved YOLOv5s, which indicates that due to the single scale and low information richness in

ML classification, artificially constructed features cannot directly determine the category of the kernel. This also confirms the advantages of DL networks in autonomous learning, optimal fusion of multi-scale features, and accurate and efficient detection. Furthermore, the detection speed of the color threshold segmentation and HOG + SVM detectors (1~2 fps) is much lower than the improved YOLOv5s (>30 fps). Therefore, it can be concluded that improved YOLOv5s is the model with the best detection performance among the three methods.

6.3.5 Effect of Different Models on Honeysuckle Detection

By gathering RGB photos and target identification algorithms, this study explored efficient techniques for field honeysuckle detection using machine vision. ML, DL, and image processing-based target identification techniques were investigated to increase the precision of image-based honeysuckle detection. Among these, the YOLOv5s model's detection accuracy in DL was 89.68%, better than that of HOG, SVM, and color threshold segmentation. There was a minimum 20% boost in the detection accuracy.

The three modules of the tiny target detection layer, the C2f module, and the Focal-EIoU were combined to enhance the YOLOv5s model and further boost the model detection efficiency. The detection performance of each method was compared, and the best detection model was selected. The results showed that the AP@0.5 of the improved YOLOv5s model in DL was 86.88%, which was 8.38% higher than the original model. The comprehensive performance index is the highest among all models. The detection speed was more than 30 fps, which met the requirements of real-time detection.

6.4 Conclusion

In response to the technical needs of the picking robot vision system for honeysuckle detection, this study successfully detected honeysuckle by constructing an RGB image data set and using three target detection methods: image processing, ML, and DL. The color threshold segmentation, HOG + SVM, and YOLOv5s were the most satisfactory models among image processing, ML, and DL technologies, with detection accuracy of 65.47%, 44.43%, and 89.68%, respectively. In addition, the comprehensive performance of improved YOLOv5s was obtained by combining the small target detection layer, C2f module, and Focal-EIoU, with P, R, AP@0.5, and F_1 values of 88.97, 79.80, 86.88, and 84.14%, respectively. The P value was reduced by 0.71% compared to the original YOLOv5s model, but other indicators were improved by more than 5%. The model is highly robust to images with different illumination, surface occlusion, and scales. The average detection speed was greater than 30 fps, which can meet real-time requirements. Therefore, the improved YOLOv5s model is recommended for a real-time field honeysuckle detection. The model needs to be embed into the honeysuckle picking robot vision system for target recognition and positioning, and tested the detection effect during actual operations. Based on the target recognition effect of field honeysuckle-picking robots, further efforts can be devoted to improve the robustness and versatility of the model.

References

1. Yan K, Cui M, Zhao S, Chen X, Tang X (2016) Salinity stress is beneficial to the accumulation of chlorogenic acids in honeysuckle (Lonicera japonica Thunb.). Front Plant Sci 7. https://doi.org/10.3389/fpls.2016.01563
2. Zhang X, Li Y, Yue C, Liu X (2023) Green and efficient cultivation techniques for honeysuckle. Seed Sci Technol 41, 49–51+66. https://doi.org/10.19904/j.cnki.cn14-1160/s.2023.24.016
3. Li X (2022) Cultivation techniques and development models of honeysuckle in Xingtai, Hebei province. Agric Eng Technol 42, 80+82. https://doi.org/10.16815/j.cnki.11-5436/s.2022.05.050
4. Liu X, Zhang Z, Igathinathane C, Flores P, Zhang M, Li H, Han X, Ha T, Ampatzidis Y, Kim H-J (2024) Infield corn kernel detection using image processing, machine learning, and deep learning methodologies under natural lighting. Expert Syst Appl 238:122278. https://doi.org/10.1016/j.eswa.2023.122278
5. Zheng R, Sun Q, Xiao G (2022) Research on honeysuckle image recognition processing algorithm based on machine vision. J Chin Agric Mech 43, 153–159. https://doi.org/10.13733/j.jcam.issn.2095-5553.2022.04.022
6. Zhang Z, Flores P, Igathinathane C, Naik L, D., Kiran, R., Ransom, J.K., (2020) Wheat lodging detection from UAS imagery using machine learning algorithms. Remote Sens 12:1838. https://doi.org/10.3390/rs12111838
7. Fernandez-Gallego JA, Lootens P, Borra-Serrano I, Derycke V, Haesaert G, Roldán-Ruiz I, Araus JL, Kefauver SC (2020) Automatic wheat ear counting using machine learning based on RGB UAV imagery. Plant J 103:1603–1613. https://doi.org/10.1111/tpj.14799
8. Flores P, Zhang Z, Igathinathane C, Jithin M, Naik D, Stenger J, Ransom J, Kiran R (2021) Distinguishing seedling volunteer corn from soybean through greenhouse color, color-infrared, and fused images using machine and deep learning. Ind Crops Prod 161:113223. https://doi.org/10.1016/j.indcrop.2020.113223
9. Zhang Z, Flores P, Friskop A, Liu Z, Igathinathane C, Han X, Kim HJ, Jahan N, Mathew J, Shreya S (2022) Enhancing wheat disease diagnosis in a greenhouse using image deep features and parallel feature fusion. Front Plant Sci 13:834447. https://doi.org/10.3389/fpls.2022.834447
10. Ren S, He K, Girshick R, Sun J (2017) Faster R-CNN: Towards real-time object detection with region proposal networks. IEEE Trans Pattern Anal Mach Intell 39:1137–1149. https://doi.org/10.1109/TPAMI.2016.2577031
11. Cai Z, Vasconcelos N (2018) Cascade R-CNN: Delving into high quality object detection. In: 2018 IEEE/CVF Conference on Computer Vision and Pattern Recognition. Presented at the 2018 IEEE/CVF Conference on Computer Vision and Pattern Recognition (CVPR), IEEE, Salt Lake City, UT, pp 6154–6162. https://doi.org/10.1109/CVPR.2018.00644
12. He K, Zhang X, Ren S, Sun J (2015) Spatial pyramid pooling in deep convolutional networks for visual recognition. IEEE Trans Pattern Anal Mach Intell 37:1904–1916. https://doi.org/10.1109/TPAMI.2015.2389824
13. Liu W, Anguelov D, Erhan D, Szegedy C, Reed S, Fu C-Y, Berg AC (2016) SSD: Single shot multibox detector. pp 21–37. https://doi.org/10.1007/978-3-319-46448-0_2

14. Lin T-Y, Goyal P, Girshick R, He K, Dollar P (2017b) Focal loss for dense object detection, in: 2017 IEEE International Conference on Computer Vision (ICCV). Presented at the 2017 IEEE International Conference on Computer Vision (ICCV), IEEE, Venice, pp 2999–3007. https://doi.org/10.1109/ICCV.2017.324
15. Redmon J, Divvala S, Girshick R, Farhadi A (2016) You Only Look Once: Unified, real-time object detection. In: 2016 IEEE Conference on Computer Vision and Pattern Recognition (CVPR). Presented at the 2016 IEEE Conference on Computer Vision and Pattern Recognition (CVPR), IEEE, Las Vegas, NV, USA, pp 779–788. https://doi.org/10.1109/CVPR.2016.91.
16. Bochkovskiy A, Wang C-Y, Liao H-YM (2020) YOLOv4: Optimal Speed and Accuracy of Object Detection
17. Wu D, Lv S, Jiang M, Song H (2020) Using channel pruning-based YOLO v4 deep learning algorithm for the real-time and accurate detection of apple flowers in natural environments. Comput Electron Agric 178:105742. https://doi.org/10.1016/j.compag.2020.105742
18. Huang R, Pedoeem J, Chen C (2018) YOLO-LITE: A real-time object detection algorithm optimized for non-GPU computers. In: 2018 IEEE International Conference on Big Data (Big Data). Presented at the 2018 IEEE International Conference on Big Data (Big Data), IEEE, Seattle, WA, USA, pp 2503–2510. https://doi.org/10.1109/BigData.2018.8621865
19. Fu L, Feng Y, Wu J, Liu Z, Gao F, Majeed Y, Al-Mallahi A, Zhang Q, Li R, Cui Y (2021) Fast and accurate detection of kiwifruit in orchard using improved YOLOv3-tiny model. Precis Agric 22:754–776. https://doi.org/10.1007/s11119-020-09754-y
20. Kuznetsova A, Maleva T, Soloviev V (2020) Using YOLOv3 algorithm with pre- and post-processing for apple detection in fruit-harvesting robot. Agronomy 10:1016. https://doi.org/10.3390/agronomy10071016
21. Wang D, He D (2021) Channel pruned YOLO V5s-based deep learning approach for rapid and accurate apple fruitlet detection before fruit thinning. Biosyst Eng 210:271–281. https://doi.org/10.1016/j.biosystemseng.2021.08.015
22. Du W, Zhu Y, Li S, Liu P (2022) Spikelets detection of table grape before thinning based on improved YOLOV5s and Kmeans under the complex environment. Comput Electron Agric 203:107432. https://doi.org/10.1016/j.compag.2022.107432
23. Freund Y, Schapire RE (1995) A desicion-theoretic generalization of on-line learning and an application to boosting. In: Vitányi P (Ed), Computational Learning Theory, Lecture Notes in Computer Science. Springer Berlin Heidelberg, Berlin, Heidelberg, pp 23–37. https://doi.org/10.1007/3-540-59119-2_166
24. Hearst MA, Dumais ST, Osuna E, Platt J, Scholkopf B (1998) Support vector machines. IEEE Intell Syst Their Appl 13:18–28. https://doi.org/10.1109/5254.708428
25. Nanni L, Brahnam S, Lumini A (2012) A simple method for improving local binary patterns by considering non-uniform patterns. Pattern Recognit 45:3844–3852. https://doi.org/10.1016/j.patcog.2012.04.007
26. Gui Z, Chen J, Li Y, Chen Z, Wu C, Dong C (2023) A lightweight tea bud detection model based on Yolov5. Comput Electron Agric 205:107636. https://doi.org/10.1016/j.compag.2023.107636
27. Lu Y, Du S, Ji Z, Yin X, Jia W (2023) ODL Net: Object detection and location network for small pears around the thinning period. Comput Electron Agric 212:108115. https://doi.org/10.1016/j.compag.2023.108115
28. Shang Y, Xu X, Jiao Y, Wang Z, Hua Z, Song H (2023) Using lightweight deep learning algorithm for real-time detection of apple flowers in natural environments. Comput Electron Agric 207:107765. https://doi.org/10.1016/j.compag.2023.107765
29. Lin Y, Chen T, Liu S, Cai Y, Shi H, Zheng D, Lan Y, Yue X, Zhang L (2022) Quick and accurate monitoring peanut seedlings emergence rate through UAV video and deep learning. Comput Electron Agric 197:106938. https://doi.org/10.1016/j.compag.2022.106938
30. Lin T-Y, Dollar P, Girshick R, He K, Hariharan B, Belongie S (2017a) Feature pyramid networks for object detection. In: 2017 IEEE Conference on Computer Vision and Pattern Recognition (CVPR). Presented at the 2017 IEEE Conference on Computer Vision and Pattern Recognition (CVPR), IEEE, Honolulu, HI, pp 936–944. https://doi.org/10.1109/CVPR.2017.106

31. Liu S, Qi L, Qin H, Shi J, Jia J (2018) Path aggregation network for instance segmentation In: 2018 IEEE/CVF Conference on Computer Vision and Pattern Recognition. Presented at the 2018 IEEE/CVF Conference on Computer Vision and Pattern Recognition (CVPR), IEEE, Salt Lake City, UT, pp 8759–8768. https://doi.org/10.1109/CVPR.2018.00913.
32. Lin T-Y, Maire M, Belongie S, Hays J, Perona P, Ramanan D, Dollár P, Zitnick CL (2014) Microsoft COCO: common objects in context. In: Fleet D, Pajdla T, Schiele B, Tuytelaars T (Eds), Computer Vision—ECCV 2014, Lecture Notes in Computer Science. Springer International Publishing, Cham, pp 740–755. https://doi.org/10.1007/978-3-319-10602-1_48
33. Yan X, Li X (2023) Research on small target detection algorithm based on improved YOLOv5, In: 2023 19th International Conference on Natural Computation, Fuzzy Systems and Knowledge Discovery (ICNC-FSKD). Presented at the 2023 19th International Conference on Natural Computation, Fuzzy Systems and Knowledge Discovery (ICNC-FSKD), IEEE, Harbin, China, pp 1–6. https://doi.org/10.1109/ICNC-FSKD59587.2023.10280925
34. Zhang D-Y, Zhang W, Cheng T, Zhou X-G, Yan Z, Wu Y, Zhang G, Yang X (2023) Detection of wheat scab fungus spores utilizing the Yolov5-ECA-ASFF network structure. Comput Electron Agric 210:107953. https://doi.org/10.1016/j.compag.2023.107953

Chapter 7
Multiscale Wheat Lodging Parameter Detection Based on MobilenetV3

Jianing Long and Zhao Zhang

Abstract Wheat is a major global food crop and plays a key role in food production and food supply. Wheat collapse has serious impacts on yield and quality, and it is crucial to obtain timely and accurate information on wheat collapse parameters. The aim of this study is to detect wheat lodging parameters using UAV images and to investigate the effect of UAV flight altitude on the model classification performance. Three UAV flight altitudes (15, 45, and 91 m) were selected to acquire images of the wheat test field, and an automatic segmentation algorithm was utilized to generate datasets at different altitudes. An improved model, MobilenetV3-M, is proposed to classify different wheat lodging parameters, and the model employs Pyramidal Convolution Pyconv (Pyramidal Convolution) to enhance the feature extraction capability of images at different scales and enhances the feature expression capability of the network through the introduction of the CA (Coordinate Attention) attention mechanism and combining it with PolyFL (PolyFocal Loss) to address the impact of data imbalance on model classification performance. The results show that the improved MobilenetV3-M has the best classification performance, with an average accuracy of 86.78%, 86.39%, and 84.35% in the datasets Angle, Location, and Ratio, respectively. Comparing the machine learning method SVM (Support Vector Machine) and the deep learning methods EfficientnetV2, MobilenetV2, and MobilenetV3, MobilenetV3 performs the best at all heights, with three datasets averaging accuracies of 76.70%, 78.17%, and 74.70%, respectively. With the increase in UAV flight altitude due to the loss of image feature information, the classification performance of both machine learning and deep learning models is affected, and this study provides a technical solution for accurately obtaining the fall parameters of wheat UAV images. The proposed improved model achieves high accuracy. It provides a new solution for wheat lodging warning and crop management, which is important for improving wheat production efficiency and sustainable agricultural development.

J. Long · Z. Zhang (✉)
College of Information and Telecommunications, China Agricultural University, Beijing 100080, China
e-mail: zhaozhangcau@cau.edu.cn

Key Laboratory of Agricultural Information Acquisition Technology, Ministry of Agriculture and Rural Affairs, China Agricultural University, Beijing 100083, China

Y. Yun et al. (eds.), *Advanced Sensing and Robotics Technologies in Smart Agriculture*, Smart Agriculture 9, https://doi.org/10.1007/978-981-97-6441-9_7

Keywords Image processing · Deep learning · Machine learning · Wheat lodging · Feature fusion

7.1 Introduction

As one of the world's most important food crops, wheat yield and quality are of great practical and strategic significance in ensuring national food security and sustainable agricultural development. Crop failure is one of the main causes that affects wheat yield and reduces its quality [1, 2]. It is usually defined as stem failure, where the stem is bent or displaced from its upright state, and root failure, where the root-soil attachment is damaged. For a long time, wheat has been prone to stem collapse due to natural disasters such as heavy rainfall, typhoons, pests, and diseases, while excessive nitrogen fertilization will lead to root collapse and reduce crop resilience, as well as have some negative impacts on the environment [3–5]. Studies have shown that different lodging parameters (lodging position, lodging angle, and lodging rate) have different degrees of influence on wheat yield. Among them, the earlier the time of the fall of wheat, the greater the fall angle and fall rate, which have a more serious impact on wheat yield and quality [6]. For example, a fall angle of 45° resulted in an 18% [7] loss in wheat yield. Meanwhile, a fall angle of 80° will result in wheat yield losses of 7–35% [8], 4–36% [9], and 54% [7]. Therefore, in order to ensure wheat yield and quality, it is crucial to monitor and evaluate wheat lodging out parameters in real time. The widely used and accepted method for assessing wheat lodging out parameters is manual field assessment, but manual evaluation methods lack objectivity and accuracy and are time-consuming and laborious. With the development of computer technology, visual monitoring, as a low-cost, accurate, rapid, and objective method of detecting crop collapse parameters, is of great significance for real-time monitoring and assessment of wheat collapse parameters, which provides key support for improving wheat yield and variety selection.

At present, the widely used and accepted evaluation method of wheat lodging parameters is manual field evaluation, but the manual evaluation method lacks objectivity and accuracy and is time-consuming and laborious [10]. With the development of computer technology, the visual detection method, as a low-cost, accurate, rapid, and objective method of crop lodging parameter detection, is of great significance for real-time detection and evaluation of wheat lodging parameters and provides key support for improving wheat yield and variety breeding.

In recent years, with the rapid development of sensor technology and artificial intelligence, near-earth remote sensing technology equipped with intelligent detection algorithms has been widely used in the agricultural field. Compared with the high cost of satellite technology and millimeter wave radar technology, drones equipped with RGB cameras become the most cost-effective choice because of their high mobility, real-time, and low cost. High-quality drone images need to develop suitable algorithms for application in the agricultural field. According to the method of feature extraction, algorithms are usually divided into traditional machine learning and deep learning methods [11, 12]. When extracting features in traditional machine learning, it is usually necessary to manually select or design the features and then input the extracted features into the machine learning classifier. For example, Rajapaksa et al. (2018) trained support vector machine (SVM) classifiers by extracting grayscale co-occurrence matrix (GLCM) and local binary mode (LBP) of images, respectively, to achieve lodging prediction of wheat and rape. [13] enhanced texture images by YCBCR transform, analyzed the influence of different bands on wheat lodging classification, and applied SVM to the classification of wheat lodging by color features and G1 band texture features enhanced by YCBCR transform. Similarly, [14] combined the spectral and texture features of UAV images to describe the lodging region of wheat with an accuracy greater than 80%. The results show that the traditional machine learning method can realize the application of wheat lodging detection. However, the above studies only classified wheat lodging under the condition of single characteristics and lacked the influence of different characteristics and combination methods on the classification effect.

Although the traditional machine learning method has made great progress in wheat lodging detection, it is still a challenge for wheat lodging detection under practical conditions due to different planting methods and the complexity of UAV images. Especially in the face of small plots of wheat field size inconsistency, poor image quality, and other conditions, the traditional machine learning manual feature extraction method cannot accurately complete the lodging detection of wheat. The advantage of deep learning is to learn the high-level hierarchical structure features of images from the training data and extract the abstract features that are closer to the high-level semantics of images [2]. Extracted five features of drone wheat images and trained three machine learning classification models (random forest (RF), convolutional neural network (CNN), and support vector machine (SVM), respectively) to classify the fallen wheat on different dates. The classification results showed that the data collected on different dates did not affect the classification performance of the model. GoogLeNet has an average accuracy of 93%. [15] adopted the image semantic segmentation model of a fully connected network (FCN-AlexNet) and SegNet neural network to identify rice bedding in UAV images, and the F1 scores reached 0.80 and 0.79 [16]. On the basis of the SegNet network, jump connection, separable convolution, and conditional random field improvement were added, respectively. Compared with support vector machines (SVM), full convolutional networks (FCN), and the original SegNet, the improved model achieved the best accuracy of 89.8%. However, most of the above and existing studies roughly divided wheat lodging parameters into lodging and non-lodging and did not further refine lodging types. Secondly, few studies have explored the effect of UAV flight height on wheat classification performance. This will make it difficult for wheat growers to select lodging-resistant varieties and field management.

For drone image preprocessing, the traditional method is usually for researchers to manually crop and divide the complete image into wheat data of different small plots to generate data sets [13]. However, due to the subjective consciousness of researchers, processing equipment, processing methods and other factors, the acquired image data size is uneven, the accuracy is low, and the image quality is low. Therefore, it is necessary and meaningful to design a method for automatically processing UAV images and generating small plot data sets.

In summary, the existing studies only unilaterally assessed wheat lodging, did not further subdivide the severity of wheat lodging, and lacked the verification of UAV flight height on the performance of wheat lodging parameters classified by model. Secondly, although the image classification method based on convolutional neural networks (CNN) solves the problem of the low accuracy of traditional machine vision classification, there are still some problems, such as complex network structure and insufficient use of multi-scale image feature information. In order to solve the above problems, a wheat lodging parameter detection model based on MobilenetV3 multi-scale feature fusion was proposed. The multi-scale feature extraction module (Pyconv) was designed from the perspective of a multi-scale convolution kernel using the idea of feature pyramid structure. The feature extraction was completed on wheat images at different scales by paying attention to both global feature information and local detail feature descriptions. At the same time, the extracted features were combined with the CA attention mechanism [17] to enhance the extraction ability of important features of the model, and the PolyFL with Polyloss framework [18] was introduced to solve the category imbalance problem. The model is trained on different data sets (angle, location, and ratio) and compares the classification performance of other machine learning and deep learning models. Lodging parameters are tested more comprehensively, and the effect of UAV flight height on the classification performance of different models is discussed. The main contributions of this paper are as follows:

(1) A wheat inversion parameter detection model based on MobilenetV3 multi-scale feature fusion was proposed. Multi-scale convolution was used to improve the feature extraction ability of the model, combined with the CA attention mechanism to enhance the feature expression ability, and polyFL was used to balance the influence of uneven data distribution on the model.
(2) In this study, lodging parameters were detected from multiple angles, corresponding data sets were constructed, and comparative tests were conducted on different models. The results showed that the flight height of the UAV had an impact on the classification performance of different models. The proposed model has better classification performance than other models at different heights and has higher robustness. Compared with the original model, the effect of UAV flight height on model classification performance was improved to some extent.

(3) An automatic segmentation algorithm suitable for UAV images is proposed, which can automatically correct the images acquired by the UAV and segment and label them according to different actual needs.
(4) Multi-feature fusion is adopted to improve the classification performance of SVM without increasing the calculation cost.

7.2 Materials and Methods

7.2.1 *Data Acquisition*

In this study, a DJI Phantom 4 Pro V2.0 UAV was used to collect RGB images. The photo pixel size is 5472 × 3078. Filming took place on August 23, 2020, in Thompson, North Dakota, USA (UTM WGS 84 14 N). In order to ensure the complete capture of the whole picture of the wheat test field, the minimum flight height of the UAV was set at 15 m (image resolution was 0.4 pixels/cm), the maximum height was 91 m (image resolution was 2.5 pixels/cm), and 45 m (image resolution was 1.2 pixels/cm) was selected as the intermediate reference value to collect the RGB image of the wheat test field. In the case of clear weather, the three data were collected at the same time to avoid the impact of light changes on the image data, and the flight time was about 12 to 15 min. After the UAV image collection, agronomic experts made subjective observations on the test field to obtain the real lodging parameters of the wheat test field, and the wheat lodging parameters were divided into three categories: no lodging, mild lodging, and severe lodging. In order to comprehensively evaluate the severity of wheat lodging, wheat lodging was further divided into 7 lodging parameters, including no lodging, root lodging, stem lodging, lodging angle greater than 45°, lodging angle less than 45°, lodging ratio greater than 50%, and lodging ratio less than 50%. The above results were recorded to provide reliable data for subsequent studies.

7.2.2 *Automatic Data Generation*

RGB images acquired by the UAV contain 464 long rectangular test plots (size 1.5 × 15.0 m, 116 plots, ID1) and short rectangular test plots (size 1.5 × 3.7 m, 348 plots, ID2), of which 424 plots are in the experimental group and 40 plots are in the control group. In order to obtain the wheat test plots of the experimental group according to the actual demand. This study proposes an automatic segmentation algorithm for wheat test field images acquired by UAVs. The main idea is to obtain the structural parameters of a single sample through multiple segmentations to realize sample labeling and automatic segmentation. The specific steps are as below:

(1) UAV image correction: the RGB image acquired by the UAV is used to eliminate the noise by the Wiener filter, and the binary edge image is calculated by the edge extraction operator, and then the tilt angle of the image is detected by using the Radon transform based on the Radon transform, and the image is automatically corrected.
(2) Region of interest extraction: In order to reduce the redundancy contained in the corrected image, this study first converts the RGB image to an HSV image, selects the color threshold of the region of interest according to the color distribution, and obtains the region of interest (ROI) by combining the S-channel and the V-channel with the fine-tuning method.
(3) Image segmentation: the region of interest is obtained by using the ultra-green algorithm (EXG) to obtain the green part and binarized, filling the small holes in the white area of the binarized image through image morphological processing, connecting the neighboring pixels, smoothing the boundaries, and tracking the regional boundaries of the binarized image to obtain the edge contour to determine the segmentation position. Then, the segmented image is binarized to obtain the edge contour, and the segmentation of long and short rectangular test fields is realized by traversing the Y-axis pixel points to accumulate pixel values, searching for pixel accumulation value surge points, and obtaining the values of the segmentation coordinate points (X) of the long and short rectangular test fields.
(4) Image automatic labeling and segmentation: the two kinds of rectangles obtained in the above steps are extracted from the edge contour, and the structural parameters are obtained, respectively. Subsequently, an automatic labeling algorithm is designed to label the wheat test field according to the actual needs, segment it into separate samples according to the experimental group and the control group, and save them.
(5) Data set generation: According to the observation results of the agronomic experts, 424 images obtained by the automatic segmentation algorithm were made into three data sets of Angle, Location, and Ratio, of which the Angle data set contained three categories, namely, not fallen, 202 images; fallen angle greater than 45°, 109 images; and fallen angle less than 45°, 113 images; the Location dataset contains three categories, namely, not fallen, 202 panels; root fallen, 71 panels; and stem fallen, 151 panels; and the Ratio dataset contains three categories, namely, not fallen, 202 panels; percent fallen greater than 50%, 73 panels; and percent fallen less than 50%, 149 panels.

7.2.3 SVM-Based Multi-Feature Fusion Classification Method

In this study, SVM is improved by a multi-feature fusion classification method, which determines wheat collapse parameters by extracting multiple features from wheat images. The basic idea and flow of this method. The main part consists of data preprocessing, model training, and classification prediction. First, the wheat lodging out parameter dataset is constructed by an automatic segmentation algorithm, and the image data are processed by scale transformation, image gray scaling, and data normalization. Secondly, different feature fusion strategies were used to fuse four features, namely HOG, LBP, color, and GLCM. Finally, principal component analysis (PCA) was used to reduce the redundancy of image information generated by multi-feature fusion, and different inversion parameters were predicted by SVM classification.

7.2.3.1 Multi-Feature Extraction and Fusion

Image feature extraction is a key step in the image classification task [19], and different features contain different information. When analyzing different image features, the more effective information is obtained, the more obvious the improvement in image classification performance. In the field of image processing, a single feature can only express part of the information of an image, which affects the performance of an image classifier. In order to improve this problem, this study extracts different image features (HOG, LBP, COLOR, and GLCM) separately and tries 14 feature combination fusion methods (four sets of single feature fusion, six sets of double feature fusion, two sets of triple feature fusion, and one set of quadruple feature fusion) so as to obtain more complete information about the image. Multi-feature fusion methods increase the dimensionality of image features with the increase in the number of features, and the set of features fused from different features also contains many redundant features, [20–22] which affects the performance of image classification work.

Principal Component Analysis (PCA) is a commonly used method for data reduction and feature extraction in machine learning techniques [23]. Its main purpose is to preserve important features in high-dimensional data and remove noise and redundant features, thus reducing the data dimensionality. The core idea is to construct a set of n-dimensional orthogonal features in the high-dimensional data, use the maximum variance to map the high-dimensional features to n-dimensional features, and retain the main features to achieve data dimensionality reduction. The specific steps are as follows: first, construct a new n-dimensional matrix K based on the original data, as in Eq. 7.1; then, find the covariance matrix C of the new matrix K, as in Eq. 7.2; finally, obtain the eigenvalues based on the covariance matrix C and output the eigenvectors according to the ascending order of the eigenvectors, as in Eq. 7.3.

$$\mathrm{K}=\begin{bmatrix} K_1 \\ K_2 \\ \vdots \\ K_i \end{bmatrix}=\begin{bmatrix} k_{11} & k_{12} & \cdots & k_{1j} \\ k_{21} & k_{21} & \cdots & k_{2j} \\ \vdots & \vdots & \ddots & \vdots \\ k_{j1} & k_{j2} & \cdots & k_{ij} \end{bmatrix}_{i\times j} \tag{7.1}$$

$$C=\tfrac{1}{i-1}KK^T \tag{7.2}$$

$$Cv_k = \lambda_k v_k \tag{7.3}$$

where $[K_1, K_2, \ldots K_n]$ denotes the construction of new matrix features; $[k_{11}, k_{12}, \ldots k_{ij}]$ denotes the original data features; and KT is the transpose matrix of K.

7.2.3.2 SVM Classification

In machine learning algorithms, support vector machines (SVM) are often used for classification and regression analysis tasks. Its earliest origin is from the principles of statistics, and the core idea is to find an optimal hyperplane to maximize the separation of different classes of data samples [24, 25]. SVM can be classified as linear or nonlinear. In practice, since the data is usually nonlinear, the kernel trick is utilized to map the nonlinear problem to a higher-dimensional feature space, and soft spacing is used to deal with noise and anomalies. Usually, SVM kernel functions include a linear kernel, a polynomial kernel, and a radial basis function kernel (RBF). SVM is widely used in image information extraction, pattern recognition, and data mining due to the simple and effective construction of the mathematical model structure and its excellent performance on small samples, nonlinear data, and high-dimensional data.

7.2.4 Improvement Based on MolilenetV3 Modeling

In this study, an improved model based on the MobilenetV3 multi-scale classification network (MobilenetV3-M) is proposed for classifying wheat lodging parameters in UAV images. The model borrows the idea of FPN, and the main part consists of Pyramid Convolution (Pyconv), Bottleneck, which adds a CA attention mechanism, and PyFocalLoss, which is a loss function designed for unbalanced datasets. Firstly, this study uses convolutional expansion of the input feature map channel and extracts image features by pyramidal convolution (Pyconv). Subsequently, the CA module is added to the inverse residual structure to enhance the features after deep convolution (DW). Finally, the loss function PyFocalLoss, which solves the imbalance of difficult and easy sample data, is used to progressively optimize the classifier to improve the classification effect of the model for imbalanced data and evaluate the results of the classification network.

7.2.4.1 Pyconv (Pyramidal Convolution)

Convolutional Neural Network (CNN) is a deep learning model that is widely used in image computer vision processing tasks (cited article). CNN captures local feature information from the original data through convolutional layers. The size of the convolutional kernel determines the size of the receptive field; the smaller the convolutional kernel, the less image information is acquired, which is suitable for acquiring local features. On the contrary, the larger the convolutional kernel is and the larger the receptive field is, the more image information is acquired, which is suitable for global feature extraction. However, a large convolutional kernel will lead to a surge in computation and, at the same time, reduce the computational performance. For this study, the wheat image acquired by UAV contains different sizes of experimental fields, and a single standard convolution captures the diversity of categories and variability on the scale poorly. In order to improve the performance of image classification, drawing on the idea of FPN [26], a pyramid convolutional layer consisting of four convolutional kernels of different sizes is constructed from the perspective of multi-scale convolutional kernels. The original data is convolved to obtain the feature map with the number of Li, which is used as input by the pyramid convolution module, and the input information is processed using four different-sized convolution kernels ($9 \times 9, 7 \times 7, 5 \times 5$, and 3×3) in parallel and on multiple scales without increasing the computational cost. Among them, the convolutional kernel with a smaller sensing field focuses on the detail information in the fallen wheat image and captures the small range of fallen wheat information, while the larger convolutional kernel can provide reliable details of the large range of fallen wheat information with contextual information.

In order to increase the diversity of convolutional kernel combination methods and to reduce the number of parameters and computational cost, this study adopts a bidirectional pyramid approach. By grouping convolutional ideas, the input features are divided into different groups, and separate convolutional kernels of different sizes are utilized for each group to achieve an increase in the size of the convolutional kernel from Level 1 at the bottom of the pyramid module to Level n at the top while reducing the depth of the convolutional kernel. Each input feature needs to connect all the output features; at this time, the depth of the convolution kernel is equal to the number of input features, the number of parameters of the standard convolution, and the computational cost of the formula. Pyconv divides the input feature map into four groups; the depth of the convolution is changed from 8 to 2, the number of parameters, and the computational cost are as in the following equation.

$$parameters = K_1^2 \times L_i \times L_o \tag{7.4}$$

$$parameters \; = K_n^2 \times L_{on} \times \frac{L_i}{\left(\frac{K_n^2}{K_1^2}\right)} + \cdots + K_3^2 \times L_{o3} \times \frac{L_i}{\left(\frac{K_3^2}{K_1^2}\right)} + \cdots K_1^2 \times L_{o1} \times L_i \tag{7.5}$$

$$FLOPs = parameters \times (H \times W) \tag{7.6}$$

where parameters denote the number of parameters; FLOPs denote the amount of computation; K denotes the size of the convolution kernel; $\frac{L_i}{(\frac{K_n^2}{K_1^2})}$ denotes the depth of the convolution kernel; W and H denote the length and width of the input image, respectively. L_i denotes the number of input feature maps; L_o denotes the number of output feature maps; and $L_{on} \cdots L_{o1}$ denotes the number of feature maps in different layers, and in Eq. 7.5, $L_{o1} + L_{o2} + L_{o3} + \cdots + L_{on} = L_i$ distributes the computed number of parameters and computation amount evenly to every layer, thus making the total number of parameters and computation of Pyconv approximate the standard convolution.

7.2.4.2 Attention Mechanism

Wheat images captured using UAVs in natural environments possess complex and redundant feature information, which is disruptive to the correct detection of wheat fall parameters. The attention mechanism, as one of the key modules in computer vision tasks, is commonly used to enhance the feature expression ability, filter out a small amount of important information from a large amount of information, and focus on this important information, ignoring most of the unimportant information, to improve the detection performance of the model [27]. Usually, traditional attention mechanisms only consider re-measuring the importance of each channel by modeling the channel relationship, which cannot capture attention in the spatial dimension. In order to improve the expressiveness and performance of the model, this study introduces the CA attention mechanism that can consider both the channel dimension and the spatial dimension at the same time.

The CA attention mechanism is a lightweight and efficient attention module based on parametric coordinates, which enables the module to adaptively adjust the feature map on different spatial scales by embedding the parameter information in the channel attention, thus expanding the scope of mobile network participation and realizing the enhancement of the expression ability of the learned features of the mobile network. Different from channel attention, the CA attention mechanism, in order to avoid the loss of spatial information by global pooling, globally pools the channel attention into a one-dimensional feature encoding process in both horizontal and vertical directions, as shown in the following equation:

$$z_c = \frac{1}{\mathrm{H} \times \mathrm{W}} \sum_{\mathrm{i}=1}^{\mathrm{H}} \sum_{\mathrm{j}=1}^{\mathrm{W}} \mathrm{x_c(i,j)} \tag{7.7}$$

$$z_c^h(h) = \frac{1}{\mathrm{W}} \sum_{0 \le \mathrm{i} < \mathrm{W}} x_c(\mathrm{h,i}). \tag{7.8}$$

$$z_c^w(w) = \frac{1}{H} \sum_{0 \le \mathrm{j} < \mathrm{H}} x_c(\mathrm{jwe}\ ). \tag{7.9}$$

where z_c is the output associated with channel c; $z_c^h(h)$ denotes the aggregation of features along the horizontal coordinate; $z_c^w(w)$ denotes the aggregation of features along the vertical coordinate; H denotes the height of the image; W denotes the width of the image; x denotes the given input; and c is the number of channels. By using the above approach, accurate location information is retained while ensuring that the spatial direction can capture long-range dependencies. Meanwhile, the output pair of direction-aware feature maps helps the network locate the target more accurately. In the main steps, firstly, the feature maps in the width and height directions are spliced, and the dimensionality is reduced by the convolution module, followed by batch normalization of the data into the sigmoid activation function, and finally the feature map f is obtained, as shown in the following equation:

$$f = \delta\left(F_x\left(\left[z^h, z^w\right]\right)\right) \tag{7.10}$$

F_x is the convolutional transform function; $[z^h, z^w]$ denotes a cascade operation along the spatial dimension; δ is a nonlinear activation function; and $f \in R^{\frac{C}{r} \times (H+W)}$ is an intermediate feature map encoding spatial information in the horizontal and vertical directions. Subsequently, the feature map f is transformed into a tensor with the same number of channels as the input x using the convolutional transform function in the horizontal and vertical directions, respectively, and f is divided into two independent tensors of $f^h \in R^{\frac{C}{r} \times H}$ and $f^w \in R^{\frac{C}{r} \times W}$ along the spatial dimension and inputted into the sigmoid activation function, and the final output of the attentional weight in the two directions is shown in the following equation:

$$g^h = \sigma\left(F_h\left(f^h\right)\right) \tag{7.11}$$

$$g^w = \sigma\left(F_w\left(f^w\right)\right) \tag{7.12}$$

where g^h is the horizontal direction weight; g^w is the vertical direction weight, and finally the y output from Coordinate Attention can be obtained by multiplication weighting calculation on the original feature map, combining the two direction attention weights and expanding as shown in the following equation:

$$y_c(i,j) = x_c(i,j) \times g_c^h(i) \times g_c^w(j) \tag{7.13}$$

7.2.4.3 Loss Function

The loss function in deep learning models can be used to determine the advantages and disadvantages of model training and determine the direction of model optimization. The cross-entropy (CE) loss function is widely used in image classification tasks due to its excellent performance. It is mainly used to measure the difference between the true probability distribution and the predicted probability distribution. For a single sample, the cross-entropy loss is calculated as follows:

$$L(y,p) = -\sum y_i \times \log(p_i) \tag{7.14}$$

where y_i denotes whether the i category is a real category, p_i denotes the probability of the i category. However, the cross-entropy function usually causes the model to pay more attention to the category with more samples due to the data distribution imbalance problem, which affects the training effect of the model. In order to solve the problem of category imbalance caused by positive and negative samples, (Lin et al., 2017b) propose an improved loss function Focal Loss based on binary cross entropy that dynamically reduces the weight of easy-to-distinguish samples in the training process through the introduction of a weight factor $(1-p)^{\lambda}$ so that the model pays more attention to hard-to-distinguish samples, thus making it easier to converge. Its formula expression is shown below:

$$\mathrm{FL}(p) = -\alpha(1-p)^{\lambda}\log(p) \tag{7.15}$$

where α is the category weight and λ denotes the weight of difficult samples. Focal loss effectively copes with performance problems caused by uneven data categorization without changing the original data distribution. However, it may significantly affect the stability of the model when facing extremely difficult samples [18]. Proposed a loss function framework called PolyLoss based on linear combinations of polynomial functions that is able to adjust the polynomial loss weights and reduce the difficulty of hyper-parameter adjustment with a simple strategy. In this study, an improved loss function FocalLoss based on the PolyLoss framework is used, which is able to vertically modify the polynomial coefficients and approximate the function by Taylor expansion, so that the loss function can flexibly adjust the importance of different polynomial bases based on the distribution of the dataset, which makes the model learning more balanced, prevents the model from being overly biased towards a larger number of categories, and has better generalization. The details are shown in the following equation:

$$\mathrm{L_{FL}} = (1 - \mathrm{P_t})^{\gamma}\log(\mathrm{P_t}) = \sum_{j=1}^{\infty} 1/j(1 - \mathrm{P_t})^{j+\gamma} \tag{7.16}$$

$$-\frac{d\mathrm{L_{FL}}}{d\mathrm{P}} = \sum_{j=1}^{\infty}(1+\gamma/j)(1-\mathrm{P_t})^{j+\gamma-1} = (1+\gamma)(1-\mathrm{P_t})^{\gamma} + (1+\gamma/2)(1-\mathrm{P_t})^{1+\gamma} \tag{7.17}$$

where $\alpha_j \in R^+$, denotes the weight of the polynomial loss; P_t denotes the probability of the target label prediction; L_{FL} is the Focal Loss Taylor expansion; and γ is the modulation factor.

7.2.4.4 Test Setup and Evaluation Indicators

All the network models in this study are built under Linux servers; the deep learning model is implemented based on PyTorch version 1.7.1, Python version 3.8, and CUDA 11.0; the development environment of the Python program is VScode; the hardware is used to test on the server with an NVIDIA RTX A5000 graphics card; the size of the video memory is 16 G; the CPU is an Intel (R) Xeon (R) Platinum 8358P CPU at 2.60 GHz; and MobilenetV3-small. Pre-training weights are used for model inference training. The model training hyperparameters are set as follows: batch size is set to 16, epoch is set to 100, initial term rate is 0.0001, learning rate change strategy is equal interval change with a step size of 2, and the training phase uses the Adaptive Learning Rate Approach (Root Mean Square Propagation, RMSProp) optimizer with momentum set to 0.9, alpha set to 0.9, and weight recession set to 0.00001.The model dataset is randomly divided into training, validation, and test sets in the ratio of 8:1:1. Among them, the training and validation sets were used for model training, and the test set was used for model validation.

In order to test the model's effectiveness in predicting wheat lodging in complex wheat field environments, the training model was evaluated in this study using accuracy, precision, recall, and F1-Score, as shown in Eqs. 7.18–7.21. *TP*, *TN*, *FP*, and *FN* are the true, true-negative, false-positive, and false-negative classes, respectively. Where accuracy indicates the ratio of the number of samples correctly classified by the model to the total number of samples, the higher precision indicates the higher accuracy of the model's classification. Precision indicates how many of the samples predicted by the model to be positive are true positive examples, and high accuracy indicates that the model is not prone to misclassifying negative examples as positive examples. Recall indicates the successful prediction of the model to be a positive example out of all the true examples, and a high recall indicates the model's ability to capture positive cases. F1-Score is a combined measure of precision and recall, which is a commonly used comprehensive evaluation metric for classification models. It is able to avoid a single extreme value of precision or recall and strike a balance between accuracy and recall.

$$\text{Accuracy}=\frac{TP+TN}{TP+TN+FP+FN} \tag{7.18}$$

$$\text{Precision}=\frac{TP}{TP+FP} \tag{7.19}$$

$$\text{Recall}=\frac{TP}{TP+FN} \tag{7.20}$$

$$F_1-\text{Score}=\frac{2\times \text{Pr ecision}\times \text{Recall}}{\text{Pr ecision}+\text{Recall}} \tag{7.21}$$

7.3 Results and Discussion

7.3.1 Experimental Results of Machine Learning Multi-Feature Classification

7.3.1.1 Feature Comparison Experiment

In image classification tasks, traditional machine learning often extracts single features to classify images. In this experiment, a multi-feature fusion information extraction method is used to enhance the model's ability to obtain useful information and improve SVM classification performance. Four features (HOG, LBP, color, and GLCM) are selected for the experiment, respectively, to be fused using different combinations and validated on a ratio dataset.

HOG features have the highest accuracy when SVM is used to classify in a single-feature manner. Compared with the other three features HOG features describe the image texture and shape information that effectively expresses the wheat lodging parameters, which is the most significant feature information. The COLOR + GLCM combination is optimal when two feature combinations are used, but the overall performance is 3.33% lower than the average accuracy of HOG features. This may be due to the lack of complementarity between the two feature combinations, and the redundancy of information generated by the fusion features, resulting in lower classification performance. The HOG + LBP + COLOR combination is the best among the three feature combinations, and the average accuracy increases by 0.78% compared to the single-feature HOG, and the average accuracy of the two combinations increases by 4.12%. This may be due to the fact that the three features are independent of each other, the HOG feature weights are higher, and LBP and COLOR provide better complementarity. When four features are combined in a way, the overall classification performance is optimal, with an accuracy of 74.12%, compared to single-feature HOG, two-feature, and three-feature combination ways, the average accuracy increases by 3.53, 6.86, and 2.75%, respectively. This is mainly due to the fact that the four-feature combination describes the feature information of the image more comprehensively. The experimental results show that the multi-feature fusion method proposed in this study has better classification performance compared to single feature and can be applied to wheat lodging parameters detection.

7.3.1.2 SVM Multi-Feature Classification Results

The classification results of SVM using multi-feature fusion for different altitudes (15, 45, and 91 m) under the datasets Angle, Location, and Ratio, respectively. The experimental results show that the classification performance of SVM in the three datasets shows a decreasing trend when the height of the UAV increases from 15 to 91 m. In the angle dataset, accuracy, F1-score, precision, and recall decreased by 7.05, 6.87, 8.03%, and 9.04%, respectively. In the location dataset, accuracy, F1-score, precision, and recall decreased by 8.24, 7.14, 8.12, and 6.53%, respectively. In the ratio dataset, accuracy, F1-score, precision, and recall decreased by 11.76, 13.32, 11.21, and 15.02%, respectively. The average accuracy, F1-score, precision, and recall of the three datasets were reduced by 9.02, 9.11, 9.12, and 10.20%, respectively. The main reason for the above problems is that when the UAV flies at a lower altitude, the closer it is to the wheat test field, the sensor is able to acquire a higher resolution image, which contains more information about the target details, and the extracted feature information is richer. Therefore, the model is able to acquire more useful features.

The confusion matrix puts the predicted results and actual results of all categories into the same table by category. The number of correct identifiers and the number of incorrect identifiers for each category. The diagonal line is the confidence level of the correct identification behavior; dark blue indicates high confidence, while other misidentification behaviors are shown in white, indicating low confidence. It can be concluded from the information that the recognition accuracy of the wheat unlodging category is relatively high for different heights and different data sets. This is due to the imbalance of data in this study, in which the number of unlodging data is much higher than the other two categories, and the model pays too much attention to the category with more data.

7.3.2 Comparative Experiments with Different Models

In this study, EfficientnetV2 (Tan and Le, n.d.), MobilenetV2 [28], and MobilenetV3 [29] were selected for comparative experiments to validate the classification performance of the proposed model on wheat lodging parameters. The effectiveness of the proposed method in this study is verified by comparing the training process curves of different models on the three datasets. From the training process curves, it can be concluded that the training process curves of the three comparison models are relatively close to each other, but MobilenetV3 appears to have optimal accuracy in the Angle, Location, and Ratio datasets, which are 80.00, 81.25, and 80.00%, respectively. The improved model based on MobilenetV3 obtains excellent results in the training process curve. Meanwhile, the method proposed in this study shows high generalization and optimal accuracy on the test set. The training accuracy tends to converge when the epoch is around 10.

The classification accuracy of different deep learning models in the test set. The experimental results show that the classification performance of the four deep learning models in all three datasets is affected by the flight height of the UAV. The reason is the same as the SVM analysis above. When the flight altitude is increased from 15 to 91 m, the overall analysis has the smallest impact on EfficientnetV2, with an average accuracy decrease of 5.47% for the three datasets, while MobilenetV3 has the largest impact with an average accuracy decrease of 7.21% for the three datasets.

This is due to the structural design of the model itself. EfficientnetV2 uses multi-scale feature fusion technology, which has better adaptability to objects of different scales. The improved MobilenetV3-M is relatively better than MobilenetV3 due to the addition of multi-scale convolution, and the average accuracy of the three datasets decreases by 6.83%. Analyzing the four models from a single dataset, the average accuracy decreases by 6.03, 6.85, and 6.87% for the angle, location, and ratio datasets, respectively, with the smallest impact on the angle dataset and a similar impact on location and ratio. The main reason for the above problem is due to the influence of data distribution. Deep learning models are prone to favoring the prediction of a larger number of categories when faced with an unbalanced dataset, while for a smaller number of categories, fewer category features are learned during the training process, which reduces the prediction ability. The angle data distribution is more balanced compared to the other two datasets. Comparing the other models under different heights and different datasets, the improved model, MobilenetV3-M, achieves the best classification performance in the test set. Comparing the baseline model, MobilenetV3 improves the average accuracy by 10.08, 8.22, and 9.66% in the angle, location, and ratio datasets, respectively. It also optimizes the effect of height on the model's classification performance to some extent. It demonstrates the classification results of the improved model MobilenetV3-M on wheat lodging parameters in different datasets, and it can be seen that the improved model can accurately classify different wheat lodging parameters, which is suitable for wheat lodging parameter detection.

7.3.3 Ablation Experiment

In order to verify the validity, feasibility, and optimization of the method proposed in this study, three models are selected for ablation experiments. The experimental parameters are set to be consistent with the training, and the module performance is verified in the angle, location, and ratio datasets, respectively. Model 1 is the original network; Model 2 is to replace CEloss (CrossEntropyLoss) with PolyFL (PolyFocalloss); Model 3 is to replace SE attention mechanism with CA attention mechanism; Model 4 is to replace Conv with Pyconv. The validation results in the Angle, Location, and Ratio datasets, respectively. Replacing the PolyFL loss function helps the model improve its classification performance due to the imbalance of data distribution, and the average accuracy, precision, recall, and F1-score are improved by 2.98, 5.85, 4.50, and 4.29%, respectively. On this basis, by replacing the attention mechanism with the CA attention mechanism, which possesses stronger mobile network learning feature expression capability, the model classification performance score is improved, with the average accuracy, precision, recall, and F1-score improved by 2.67, 2.23, 3.75, and 3.55%, respectively. In order to model both global and local feature extraction and to improve the feature extraction ability for images of different sizes, With the introduction of the Pyconv multiscale convolution module, the model classification performance is further improved, with average accuracy, precision, recall, and F1-score sums of 3.36, 4.97, 4.28, and 4.55%, respectively. Overall, the classification performance of the improved model is significantly improved compared to the baseline model.

In addition, to illustrate the effectiveness of PolyFL even further, this study presents the test results of using CEloss with PolyFL for each category in the Angle, Location, and Ratio datasets, respectively. The experimental results show that the average F1-score for the three categories using CEloss on the Angle, Location, and Ratio datasets is 90.06, 64.73, and 75.94%, respectively. It can be seen that the model focuses too much on one category, resulting in its F1-score value being much larger than the other two categories. While using PolyFL on Angle, Location, and Ratio datasets, the average F1-scores of the three categories are 89.41, 72.12, and 81.13%, respectively, which are closer to the F1-score values of each category compared to CEloss. The above results indicate that PolyFL can effectively improve the impact of the model due to the imbalance of data distribution.

7.3.4 Comparative Experiments on the Improvement of Different Modules of the Model

Among the deep learning models, MibilenetV3 is used as the baseline model in this study, which is improved for convolution, attention mechanism, and loss function, respectively. In order to illustrate the effectiveness of the improved modules, comparison experiments were conducted, and all experiments were unfolded under the Angle, Location, and Ratio datasets, respectively. Comparing Model 1, Model 2, and Model 3 with only the improved loss function, it can be seen that the selected improved loss function has improved the classification performance of the model to different degrees. Among them, Model 3 under each dataset has the most obvious improvement, with the overall average accuracy, precision, recall, and F1-score of the three datasets improved by 2.98, 5.85, 4.50, and 4.29%, respectively. Thereafter, different attention mechanisms (GAM and CA) are compared on the basis of the two improved loss functions. The experimental results showed that compared to the GAM attention mechanism, the CA attention mechanism was more used with the wheat collapse parameter detection in this study under the two loss functions, and compared to Model 3, Model 7 had the most obvious improvement effect, with the overall average accuracy, precision, recall, and F1-score improved by 1.92, 2.56, 3.74, and 3.41%, respectively, and the results further illustrate the validity of the selected modules in this study. Finally, adding multi-scale convolution to Model 7 optimizes the model classification performance, with the overall average accuracy, precision, recall, and F1-score of the angle, location, and ratio datasets improving by 4.11, 4.65, 4.30, and 4.69, respectively.

7.3.5 Comparison of Machine Learning and Deep Learning Models

The improved SVM and MobilenetV3-M have the best performance in detecting wheat lodging parameters in machine learning methods and deep learning methods, respectively. To further analyze the performance of the two models, this study outputs the ROC curves for each class in angle, location, and ratio data, respectivel. As can be seen from the ROC curves, the ROC curves for each class of the deep learning MobilenetV3-M model are closer to the upper left corner than SVM, indicating better classification performance. In this study, the Area Under Curve (AUC) value is used to describe the model's classification performance in different categories in more detail. MobilenetV3-M improves the average AUC by 0.11, 0.08, and 0.14 for the Angle, Location, and Ratio datasets, respectively, compared to SVM. The maximum improvement of AUC for the collapse parameter with an angle less than 45° is 0.26, and the minimum improvement for the uncollapsed parameter is 0.02 in the Angle dataset. Further analysis shows that, compared to SVM, MobilenetV3-M improved the overall average accuracy, precision, recall, and F1-score by 17.74, 22.58, 24.66, and 24.09% in the angle, location, and ratio datasets, respectively. The most significant improvement in the Angle dataset is 17.77, 23.32, 26.07, and 24.91% for average accuracy, precision, recall, and F1-score, respectively, and the most significant improvement in the Angle dataset is at a height of 91 m for MobilenetV3-M. The average accuracy, precision, recall, and F1-score are improved by 19.77, 26.47, 28.51, and 27.68%, respectively. The improvement is most obvious at 45 m height in the Location and Ratio dataset, with average accuracy, precision, recall, and F1-score improving by 18.49, 23.05, 29.01, and 27.12% versus 18.96, 18.97%, respectively, 22.23, and 20.48%. In summary, comparing the machine learning SVM classification model, since SVM classifiers usually use hand-designed features, which may not be sufficient to capture complex image features, the deep learning model MobilenetV3-M improves the classification performance significantly at lower altitudes due to the fact that photographs taken at lower altitudes when the UAV is flying are richer in feature information, and the deep learning model is more applicable to the high-dimensional data and is used to deal with the complex image features. The deep learning model is more suitable for high-dimensional data and is used to deal with data with a large number of features. Deep learning can learn and represent complex nonlinear relationships, which improves the expressive ability of feature extraction and the classification performance of the model. After comprehensively comparing the classification performance results, this study concludes that MobilenetV3-M is the best model for classifying wheat lodging parameters based on UAV images.

7.4 Conclusion

In this study, a multi-scale feature fusion network model is proposed for detecting the fall parameters of RGB images of wheat experimental fields acquired by UAVs at different flight altitudes (15, 45, and 91 m). The model uses MobilenetV3 as the baseline and enhances the feature extraction capability of the network for images of different sizes by introducing the multi-scale feature extraction convolution Pyconv module. It also adopts the CA attention mechanism and PolyFL to improve the feature expression capability of the network and its ability to cope with the distribution of unbalanced datasets. In order to verify the performance of the improved model, the UAV image is segmented using the self-designed automatic image segmentation algorithm, and the Angle, Location, and Ratio datasets are generated. The existing deep learning models are compared in the above datasets. The experimental results show that MobilenetV3-M possesses excellent performance in the detection of fall parameters, with an average accuracy rate of more than 80% at all three heights. In addition, the flight altitude of the UAV affects the performance of the deep learning model for inverted parameter detection, as evidenced by the fact that the classification accuracy of the model decreases with the rise of the flight altitude, and the highest accuracies achieved in the Angle, Location, and Ratio datasets when the UAV descends to the lowest altitude of 15 m are 89.66, 89.80, and 88.78%, respectively. MobilenetV3-M also achieves good classification performance on the three datasets when the UAV ascends to a maximum of 91 m, with average accuracies of 84.4, 82.65, and 80.61% all above 80.00%, which indicates that the proposed model has good robustness. Comparing with other models (SVM, EfficientnetV2, MobilenetV2, and MobilenetV3) at different heights, the average accuracies are all improved by more than 9.00%, proving the effectiveness of the model improvement. In summary, the model proposed in this study possesses excellent classification performance, which can satisfy the detection of wheat lodging parameters and provide a reference for wheat breeders and growers, insurance loss adjusters, and agronomists and plant physiologists.

References

1. Khobra R, Sareen S, Meena BK, Kumar A, Tiwari V, Singh GP (2019) Exploring the traits for lodging tolerance in wheat genotypes: a review. Physiol Mol Biol Plants 25:589–600. https://doi.org/10.1007/s12298-018-0629-x
2. Zhang Z, Flores P, Igathinathane C, Naik L, D., Kiran, R., Ransom, J.K., (2020) Wheat lodging detection from uas imagery using machine learning algorithms. Remote Sens 12:1838. https://doi.org/10.3390/rs12111838
3. Keikha M, Darzi- Naftchali A, Motevali A, Valipour M (2023) Effect of nitrogen management on the environmental and economic sustainability of wheat production in different climates. Agric Water Manag 276:108060. https://doi.org/10.1016/j.agwat.2022.108060

4. Park J-R, Jang Y-H, Kim E-G, Lee G-S, Kim K-M (2023) Nitrogen fertilization causes changes in agricultural characteristics and gas emissions in rice field. Sustainability 15:3336. https://doi.org/10.3390/su15043336
5. Pinthus MJ (1974) Lodging in Wheat, Barley, and Oats: The phenomenon, its causes, and preventive measures, in: Brady NC (Ed), Advances in Agronomy. Academic Press, pp 209–263. https://doi.org/10.1016/S0065-2113(08)60782-8
6. Nuttall JG, O'Leary GJ, Panozzo JF, Walker CK, Barlow KM, Fitzgerald GJ (2017) Models of grain quality in wheat—A review. Field Crops Res. Modeling crops from genotype to phenotype in a changing climate 202:136–145. https://doi.org/10.1016/j.fcr.2015.12.011
7. Berry PM, Spink J (2012) Predicting yield losses caused by lodging in wheat. Field Crops Res 137:19–26. https://doi.org/10.1016/j.fcr.2012.07.019
8. Fischer RA, Stapper M (1987) Lodging effects on high-yielding crops of irrigated semidwarf wheat. Field Crops Res 17:245–258. https://doi.org/10.1016/0378-4290(87)90038-4
9. Acreche MM, Slafer GA (2011) Lodging yield penalties as affected by breeding in Mediterranean wheats. Field Crops Res 122:40–48. https://doi.org/10.1016/j.fcr.2011.02.004
10. Li D, Song Z, Quan C, Xu X, Liu C (2021) Recent advances in image fusion technology in agriculture. Comput Electron Agric 191:106491. https://doi.org/10.1016/j.compag.2021.106491
11. Du X, Cai Y, Wang S, Zhang L (2016) Overview of deep learning. In: 2016 31st Youth Academic Annual Conference of Chinese Association of Automation (YAC). Presented at the 2016 31st Youth Academic Annual Conference of Chinese Association of Automation (YAC), pp 159–164. https://doi.org/10.1109/YAC.2016.7804882
12. Wang P, Fan E, Wang P (2021) Comparative analysis of image classification algorithms based on traditional machine learning and deep learning. Pattern Recognit. Lett. 141:61–67. https://doi.org/10.1016/j.patrec.2020.07.042
13. Ding Y, Zhao X, Zhang D, Liang D, Wang Z, Xi S, Du S (2019) Rice lodging area extraction based on ycbcr spatial and texture features. In: IGARSS 2019–2019 IEEE International Geoscience and Remote Sensing Symposium. Presented at the IGARSS 2019 - 2019 IEEE International Geoscience and Remote Sensing Symposium, pp 9228–9231. https://doi.org/10.1109/IGARSS.2019.8898862
14. Liu H, Guijun Y, Zhu H (2014) The extraction of wheat lodging area in uav's image used spectral and texture features. Appl Mech Mater 651–653:2390–2393. https://doi.org/10.4028/www.scientific.net/AMM.651-653.2390
15. Yang M-D, Tseng H-H, Hsu Y-C, Tsai HP (2020) Semantic segmentation using deep learning with vegetation indices for rice lodging identification in multi-date uav visible images. Remote Sens 12:633. https://doi.org/10.3390/rs12040633
16. Song Z, Zhang Z, Yang S, Ding D, Ning J (2020) Identifying sunflower lodging based on image fusion and deep semantic segmentation with UAV remote sensing imaging. Comput Electron Agric 179:105812. https://doi.org/10.1016/j.compag.2020.105812
17. Hou Q, Zhou D, Feng J (2021) Coordinate attention for efficient mobile network design. In: 2021 IEEE/CVF Conference on Computer Vision and Pattern Recognition (CVPR). Presented at the 2021 IEEE/CVF Conference on Computer Vision and Pattern Recognition (CVPR), IEEE, Nashville, TN, USA, pp 13708–13717. https://doi.org/10.1109/CVPR46437.2021.01350
18. Leng Z, Tan M, Liu C, Cubuk ED, Shi X, Cheng S, Anguelov D (2022) POLYLOSS: A polynomial expansion perspective of classification loss functions
19. Jiang C, Chen T, Tao G (2023) Gap image classification based on pca-svm with multiple feature fusion. In: 2023 4th International Conference on Mechatronics Technology and Intelligent Manufacturing (ICMTIM). Presented at the 2023 4th International Conference on Mechatronics Technology and Intelligent Manufacturing (ICMTIM), pp 609–613. https://doi.org/10.1109/ICMTIM58873.2023.10246511
20. Ao C, Ye X, Sakurai T, Zou Q, Yu L (2023) M5U-SVM: identification of RNA 5-methyluridine modification sites based on multi-view features of physicochemical features and distributed representation. BMC Biol 21:93. https://doi.org/10.1186/s12915-023-01596-0

21. Hasan MM, Tsukiyama S, Cho JY, Kurata H, Alam MA, Liu X, Manavalan B, Deng H-W (2022) Deepm5C: A deep-learning-based hybrid framework for identifying human RNA N5-methylcytosine sites using a stacking strategy. Mol Ther 30:2856–2867. https://doi.org/10.1016/j.ymthe.2022.05.001
22. Yan C, Li M, Ma J, Liao Y, Luo H, Wang J, Luo J (2022) A novel feature selection method based on mrmr and enhanced flower pollination algorithm for high dimensional biomedical data. Curr Bioinforma 17:133–149. https://doi.org/10.2174/1574893616666210624130124
23. Fischler MA, Bolles RC (1981) Random sample consensus: a paradigm for model fitting with applications to image analysis and automated cartography. Commun ACM 24:381–395. https://doi.org/10.1145/358669.358692
24. Cherkassky V, Ma Y (2004) Practical selection of SVM parameters and noise estimation for SVM regression. Neural Netw 17:113–126. https://doi.org/10.1016/S0893-6080(03)00169-2
25. Chua F-F, Lim T-Y, Tajuddin B, Yanuarifiani A (2022) Incorporating semi-automated approach for effective software requirements prioritization: A framework design. J. Inform. Web Eng. 1, 1–15. https://doi.org/10.33093/jiwe.2022.1.1.1
26. Lin T-Y, Dollár P, Girshick R, He K, Hariharan B, Belongie S (2017a) Feature pyramid networks for object detection. In: 2017 IEEE Conference on Computer Vision and Pattern Recognition (CVPR). Presented at the 2017 IEEE Conference on Computer Vision and Pattern Recognition (CVPR), pp 936–944. https://doi.org/10.1109/CVPR.2017.106
27. Niu Z, Zhong G, Yu H (2021) A review on the attention mechanism of deep learning. Neurocomputing 452:48–62. https://doi.org/10.1016/j.neucom.2021.03.091
28. Sandler M, Howard A, Zhu M, Zhmoginov A, Chen L-C (2018) MobileNetV2: Inverted residuals and linear bottlenecks. In: 2018 IEEE/CVF Conference on Computer Vision and Pattern Recognition. Presented at the 2018 IEEE/CVF Conference on Computer Vision and Pattern Recognition (CVPR), IEEE, Salt Lake City, UT, pp 4510–4520. https://doi.org/10.1109/CVPR.2018.00474
29. Howard A, Sandler M, Chen B, Wang W, Chen L-C, Tan M, Chu G, Vasudevan V, Zhu Y, Pang R, Adam H, Le Q (2019) Searching for MobileNetV3. In: 2019 IEEE/CVF International Conference on Computer Vision (ICCV). Presented at the 2019 IEEE/CVF International Conference on Computer Vision (ICCV), IEEE, Seoul, Korea (South), pp 1314–1324. https://doi.org/10.1109/ICCV.2019.00140

SPRINGER NATURE

GPSR Compliance

The European Union's (EU) General Product Safety Regulation (GPSR) is a set of rules that requires consumer products to be safe and our obligations to ensure this.

If you have any concerns about our products, you can contact us on ProductSafety@springernature.com

In case Publisher is established outside the EU, the EU authorized representative is:

Springer Nature Customer Service Center GmbH
Europaplatz 3
69115 Heidelberg, Germany

The manufacturer's authorised representative in the EU is Springer Nature Customer Service Centre GmbH, Europaplatz 3, 69115 Heidelberg, Germany. If you have any concerns regarding our products, please contact ProductSafety@springernature.com

Printed and bound by CPI Group (UK) Ltd, Croydon, CR0 4YY
01/10/2026
02225987-0009